Paweł Machcewicz
The War That Never Ends

Public History in International Perspective

Theory, Method, and Public Practice

Edited by
Indira Chowdhury and Michael M. Frisch

Volume 2

Paweł Machcewicz

The War That Never Ends

The Museum of the Second World War in Gdańsk

translated by Anna Połapska Adamek
(in cooperation with Bryan Dewalt
and David Monaghan)

DE GRUYTER
OLDENBOURG

ISBN 978-3-11-076349-2
e-ISBN (PDF) 978-3-11-065909-2
e-ISBN (EPUB) 978-3-11-065503-2
ISSN 2626-1774

Library of Congress Control Number: 2019945644

Bibliographic information published by the Deutsche Nationalbibliothek
The Deutsche Nationalbibliothek lists this publication in the Deutsche Nationalbibliografie;
detailed bibliographic data are available on the Internet at http://dnb.dnb.de.

© 2021 Walter de Gruyter GmbH, Berlin/Boston
This volume is text- and page-identical with the hardback published in 2019.
Image Credit: Michal Fludra / NurPhoto / Getty Images
Pictured on the cover is a demonstration in front of the Museum of the Second World War
in defense of its independence. Gdańsk, April 5, 2017.
Druck und Bindung: CPI books GmbH, Leck

www.degruyter.com

Preface

Treasure boxes. Memory palaces. Temples of learning. Heritage sites. Must-see tourist destinations. Museums have been described as all of these. Surveys conducted across the globe by public historians and museum professionals show that publics consider museums to be extremely trustworthy custodians of culture, history, and memory. History museums not only collect and conserve the past but tell stories about it through their architecture, displays, interactives, tours, events, cafés, and shops. Visitors come anticipating the familiar but also expecting to learn something new, even to be challenged. They want to enjoy the museum's exhibits but also to engage with them, even participate in ways that help them shape the stories being told, turning its spaces into meaningful places. Museums present us to ourselves and to others. They are contact zones where people who might otherwise never encounter each other meet and experience something together. Museums matter. This is why, in times of war, we scramble to protect them and their contents. It is why, in times of war, they become targets, their destruction symbolic of victory. We grieve when they are destroyed. And this is why, even in peacetime, they can become sites of intense controversy and even conflict.

The International Federation for Public History's steering committee believes that Paweł Machcewicz's book on the project, discussion, development, and redefinition of the Museum of the Second World War in Gdańsk offers a very rich example of contested public history that transcends national barriers. Since the opening of the Caen Memorial Museum (France) in 1988, the Museum in Gdańsk is the first major museum devoted to the Second World War that opened its doors in Europe. Moreover, it is the first one located in Eastern Europe. In both cases (Caen and Gdańsk), the cities were deeply impacted by the war. In both cases civilians and soldiers paid a high price in the conflict. The history of the Second World War remains a difficult story with entangled, contradictory, and contested narratives and traumatic experiences. Because of this, museums become negotiated places where the story is told. They are constantly striving for a difficult balance between evolving demands and expectations shaped within national frameworks on the one hand and the integration of transnational perspectives on the other. With its calls to the public for artifacts and materials to create new collections and work in the context of studies of public memory in Poland, public debates in newspapers and other media, and the involvement of historians, the Museum is an excellent example of both the very rich benefits and considerable challenges of public history practice.

Museums are, for political authorities, not only a way to represent, visualize, and exhibit the past but also a convenient medium to encourage social cohesion and promote communal and national identity. Both powerful and vulnerable spaces, museums are symbolic of public history. The experience of Paweł Machcewicz, historian, founder, and dismissed director, is an excellent example of this contradictory situation. His book presents a fascinating story of how museums are created, from the very early stages of conceptual planning, through detailed decisions made on many individual elements, to the final installations carried out under the pressure of time before the official opening. As such, the book will be very useful for public history practitioners, professionals, academics, and students all over the world. The book also offers a very detailed journey through the relationship between public history and politics as well as the tensions we see across the globe between national(istic) policies and more open, transnational contexts. Machcewicz explores how changing political regimes and policy in the present affect representations and interpretations of the past.

Since the opening of the Museum, several changes have been introduced to the main exhibition. The most noticeable is the replacement of the film that was presented at the end of the exhibition, portraying life after the Second World War on both sides of the Iron Curtain (western on the left-hand side of the screen and eastern on the right-hand side). Joined together after the collapse of the communist regime, the film also addressed contemporary issues, such as the immigration crisis. Now, visitors are greeted with an animated film, *The Unconquered*, a production of the Institute of National Remembrance, praising Polish suffering and achievements during the Second World War.

Other steps in the same direction include moving the stories of the Polish heroes of the Second World War, such as Witold Pilecki, Father Maksymilian Kolbe, and Irena Sendler, to more prominent places in the museum and adding the story of Józef and Wiktoria Ulma, who were executed with their children for their efforts to rescue Jews. Each of the 316 Polish paratroopers (*Cichociemni*) trained in Britain and dropped into the Polish territories under occupation are now listed by name. The numbers of the Polish victims and Polish participants in the armed anti-Nazi underground were seen as underestimated and have therefore been reconsidered. Last but not least, more examples of Stalinist terror were added in the part of the exhibition dedicated to the prewar Soviet Union in order to emphasize the coercive, not just propagandist, nature of the system. The authors of the first exhibition sued the new director for introducing those changes and, thus, violating copyright laws.

Of course, in the European context, the Second World War and its legacy is a very sensitive topic. But it is not the only one. Let's think, for example, of colonial and postcolonial narratives. But other topics also appear to be very sensitive: civil wars, dictatorships, and postdictatorships. How can societies remember difficult pasts? How can they rebuild social cohesion and create socially acceptable narratives? Museums appear more and more as the ideal instrument to do so if one wants to reach a broader audience. Museums are educative tools but also a place to stimulate discussion, and public debate is the essence of democratic society. But in the case of the Gdańsk museum, the possibility for debate was cut off, provoking much discussion not only in Poland but in the global museum and historical world. This is why Paweł Machcewicz's book is so important. Because history matters.

Thomas Cauvin, David Dean, Tanya Evans, Chantal Kesteloot, Anita Lucchesi, Catalina Muñoz, and Joanna Wojdon

> Steering Committee, International Federation for Public History
> April 2019

Contents

Paweł Is a Dead Man —— 1

I The Beginnings

It All Started with an Article ... —— 7

Museum, but What Museum? —— 13

"Disintegration of the Polish Nation" —— 16

History Museums: Between Collective Imagination and Politics —— 27

II Making the Museum

The First Months at the Office of the Prime Minister —— 37

Assembling the Gdańsk Team —— 48

Architecture, Archeology, and Construction —— 55

Building the Collection and Planning the Exhibitions —— 62

The Academic Advisory Committee, or a Multitude of Perspectives —— 76

From the Concept to the Exhibitions and Everything in Between —— 83

Before the Storm —— 92

III The War

"The Polish Point of View" —— 97

Minister Gliński Comes at Night —— 104

The Race Against Time —— 111

Parliamentary Committee on Culture: Hunting for "Encyclopedists" —— 116

Reviewers Unmasked —— 125

Stalingrad Instead of Blitzkrieg —— 138

The Finale —— 154

Not the End of the Story —— 173

Bibliography —— 180

Index of Names —— 182

Photos (by Bartosz Makowski): —— 185

Paweł Is a Dead Man

"Paweł is a dead man." In November 2015, ten days after the Law and Justice Party[1] formed the government in Poland, this short message was passed on to one of my colleagues at the Museum of the Second World War. It came from a mutual acquaintance, who was close to the new ruling party. He heard it, nice and clear, from the party's leadership. He also said that he had seen a list of staff to be fired from the Museum and that there was no chance that any of us would last.

The next few months were full of equally strong language and actions directed toward me and the Museum that I was creating. This was not entirely surprising since Law and Justice and its supporters had thrown accusations at us from the very beginning. The party's leader and the real ruler of Poland, Jarosław Kaczyński, had accused us of lacking "the Polish point of view" as far back as 2008. He had said that the Museum of the Second World War was a tool to "disintegrate the Polish nation." Others were saying that we were under directions from Brussels and Berlin.

Still, I could never have predicted the range of tactics that the government of my own country would employ against the largest historical museum being created in Poland in order to prevent its opening to the public and to change its exhibitions. I did expect to be fired, but I did not foresee that the minister of culture would create another, fictional museum that existed only on paper to extinguish the Museum of the Second World War. Nor did I think that one Friday night in April 2016 he would unexpectedly announce that our Museum would be merged with this nonexistent Museum of Westerplatte and the War of 1939. Not that long ago, I would have thought that this type of story could not really happen, that it belonged to the grotesque novels and dramas of Sławomir Mrożek.[2] I did not foresee that the Museum would become one of the most important public issues in Poland, a symbol of the defense of history and culture from political interference.

---

[1] Law and Justice (Prawo i Sprawiedliwość) is a conservative party with strong nationalistic, conservative, and Catholic leanings. It was formed in 2001 by brothers Lech and Jarosław Kaczyński. After Lech Kaczyński's death in an airplane crash near Smoleńsk in 2010, Jarosław became the sole leader of the party. Law and Justice won elections and formed governments in Poland from 2005 to 2007 and from 2015 to the present.
[2] Sławomir Mrożek (1930–2013) was a Polish writer and dramaturg, known for his satirical and absurdist style.

Yet, unlike so many other events that he was able to bend to his will, the story did not unfold entirely as Jarosław Kaczyński wished. We were not immediately squashed by the government and political forces, as they wanted. We mounted an opposition; we defended the exhibitions and tried to get the Museum open to the public. If I may use a wartime metaphor, this was Stalingrad, not the blitzkrieg the government had expected. Public opinion was on our side, and a court halted the closure for several months in response to lawsuits filed by me, Ombudsman for the Rights of Citizens Adam Bodnar, and Mayor of Gdańsk Paweł Adamowicz. Deputy Prime Minister and Minister of Culture Piotr Gliński appealed to a higher court, and the decision to close us loomed.

It was a race against time to complete as much as we could: to finish construction and begin installing exhibitions in such a manner that any future changes would be difficult to make. We were working under extreme pressure, under constant attack and threat from Minister Gliński and the whole state apparatus. Because I was stubborn, I became public enemy number one, and the politicians and the party's propaganda accused me of "lacking national sensibility," of "cosmopolitanism," and of conducting "German historical politics." As a historian, I was reminded of the Communist era in Poland, of the events of March 1968, and of "anti-Zionism."[3] All these charges were levied, then as now, by people who claimed to act for the good of Poland. I was also accused of misusing construction funds, an old and well-known authoritarian trick. Minister of Culture Gliński employed many of the resources at his disposal to find a "hook" with which to compromise me and my coworkers. Despite all this, they did not succeed.

For many people, our fight for the Museum was a source of hope that it was possible to successfully defend values and that we do not have to give up in the face of overwhelming forces. During these several months of work under extreme pressure, I did not believe that we could win and open the Museum to the public. But I did believe that we had to work to the end and do all we could. Still, I did not envision that I would, in fact, last until March 2017, when the Museum opened. I did not envision that I would lead in the very first visitor, ninety-six-year-old Joanna Muszkowska-Penson, a courier for the Union of Armed Struggle, a prisoner of the infamous Pawiak prison and the Ravensbrück concentration camp, and in the 1980s an active member of the Solidarity movement. For me, she was the symbol of all that is best about Poland. These are the people

---

3 A series of major anti-Communist protests organized by students and intellectuals took place in Poland in March 1968. The protests, brutally suppressed by the Security Forces, were followed by a wave of antisemitism, encouraged by the Communist government, which led to a mass exile of people of Jewish descent from Poland in the late 1960s and the early 1970s.

whom the Museum featured, a fact that cannot be changed by all the lies that have been thrown at it publicly.

This moment of triumph was shortly followed by a complete change in fortune, which was typical of the roller coaster of the preceding few months. Two weeks after the opening, in April 2017, the Museum of the Second World War was finally merged with the Museum of Westerplatte. Proclaiming that the case was not within the jurisdiction of the courts, the Supreme Administrative Court declined to issue a ruling, which allowed Minister of Culture Piotr Gliński to implement his plan. The merger that started on a Friday night a year earlier was now completed. Our institution was formally abolished, stricken from the registry of museums. Nobody even had to fire me because the institution of which I was the director just disappeared. The minister of culture replaced it with a new museum, under the same name but now staffed with his own people, who were directed to change the exhibitions we had created. Yet what we accomplished cannot be erased. The Museum can no longer be simply closed, and the exhibitions have already been visited by hundreds of thousands of people. Any changes, implemented at the direction of this government, happen in an open forum and are carefully watched by the general public.

I think it is worth telling the story from the very beginning to show how the Museum of the Second World War came to be, how we created the exhibitions, and how we managed—despite the government—to open it to the public. It is a story even more worth telling because it is not possible anymore to visit the Museum as it was planned. Some parts of the original exhibitions have already been changed, and this book can serve as a witness of all that happened.

I The Beginnings

It All Started with an Article ...

It all started with an article I published in November 2007, in *Gazeta Wyborcza*, the largest liberal daily newspaper in Poland, about Polish-German disputes over history, which had intensified in recent years.[1] These centered on the evaluation of postwar expulsions of Germans from Polish territory. In Poland, the activities of Erika Steinbach and the Union of Expellees had been highly criticized. This was part of a wider issue that Poles perceived as excessive—namely, the growing emphasis among sectors of German public opinion on the sufferings of Germans during and immediately after the war. This growing emphasis was evident in the story of German refugees fleeing the Red Army on the ship *Wilhelm Gustloff*, the event retold in the novel *Crabwalk* by Günter Grass, as well as in a popular movie made by German public-service broadcaster ZDF. It was also evident in an emphasis on Allied bombings and on postwar expulsions. In Poland there was a widespread fear that it could lead to distortions of history, presenting Germans more as victims of the war than as perpetrators. The greatest reaction in Poland was triggered by the topic of the postwar expulsions, which were to be featured in a new museum in Berlin dedicated to this theme and supported by the Union of Expellees. In this case, a broad, cross-party societal consensus prevailed in Poland, unique in the country as it became increasingly polarized politically and ideologically. "For many years," I wrote in *Gazeta Wyborcza*,

> the majority of voices in Poland questioned the legitimacy of characterizing the forced displacements as one of the most important problems in the twentieth-century history of Europe. Accepting that premise would divorce the forced displacements from the context of the Second World War and German aggression and crimes, without which resettlement would not have existed at all. It is worth remembering that in the depictions of many German historians and media, even those who distance themselves from Steinbach, expulsions of the Germans are included in the series of ethnic cleansings and forced displacements that began in Europe after the Balkan Wars of 1912–13 and to which a contemporary analogy would be the rapes and expulsions during the breakup of Yugoslavia. In this view, the main evil is the nation state, which eliminates minorities, and the common denominator for the evaluation and rejection of these actions is human rights in their modern sense. According to this logic, in expelling the Germans the Poles pursued aspirations to create a unified nation state. They were victims of Nazi crimes but also perpetrators of the suffering of innocent Germans—just like the Serbs committing a crime against Bosnians or Albanians. . . . Such a perspective deforms the actual picture of history. Moreover, it relativizes—not directly, but through its assumed vision of the twentieth century—the uniqueness of both the Second World War and of the totalitarianisms: their total evil and crimes, which are impossible

---

[1] Paweł Machcewicz, "Muzeum zamiast zasieków," *Gazeta Wyborcza*, November 8, 2007.

to logically integrate into any historical narratives that might also incorporate the Balkan wars, the disintegration of Yugoslavia, or, for that matter, any other broadly and widely criticized oppressions imposed by a nation-state on its minorities.

In my article, I proposed that Polish public opinion not limit its efforts to barren protests, which in any event would not prevent the creation of the planned Berlin museum and educational center dedicated to the expulsions. I argued that a better solution would be to present our own narrative focused on the Second World War and its consequences. Growing international disputes, including Polish-German and Polish-Russian ones, over how to interpret the events of 1939 to 1945 and their consequences clearly showed that the Second World War remained the central experience for European nations. Without it, understanding the postwar decades and even contemporary events was impossible. The best solution, I argued in the same text, would be to create in Poland the Museum of the Second World War, which would show Europe and the world the entire experience of the Second World War, with particular emphasis on the fate of Poland and other Central and Eastern European countries, a history generally little known in the western part of the continent and in the United States. This would of course have much deeper meaning beyond the Polish-German historical disputes. The new Museum would speak not only to Poles and Germans but to all Europeans and potentially also to visitors from other continents. I had already formulated this argument publicly in previous years. I proposed to expand the historical memory of Western Europe, for decades isolated from the east by the Iron Curtain and little interested in the complicated histories of nations left under Soviet rule after the war, their fates not fitting into Western Europeans' ideas about this greatest conflict in history. "It's time for a Polish initiative, tailored to a place that we could have occupied in the European Union," I argued in *Gazeta Wyborcza*, "if we had not condemned ourselves to this isolation, had not locked ourselves in a besieged fortress. . . . In such a museum, which still does not exist in Europe, there would be space to show the full experience of war, including the perspective of nations that experienced totalitarianism, not only Nazi totalitarianism but also Soviet."

My article in *Gazeta Wyborcza* appeared a few weeks after the 2007 parliamentary elections and a change in government in Poland. I counted on the possibility of new openness, of change in many areas, both in thinking about how to increase historical consciousness (I never liked the term "historical politics" as inseparably connected with the use and misuse of history) and in attempting to end the impasse in Polish-German relations, which had been so negatively affected in the recent disputes about history. In the article, I wrote that Donald Tusk, the new prime minister, could announce the creation of the Museum of the Second World War "during his first visits to the European capitals," making it immediately the subject of a wide debate. However, I did not expect such a quick and positive reaction. To be honest, I assumed that there would be no reaction. I had never met Donald Tusk. I was sure that after taking power, he would be preoccupied with more important issues than creating a museum. Yet soon after the article was published, Wojciech Duda, editor in chief of the respected magazine *Przegląd Polityczny* and an advisor to Donald Tusk, asked me to elaborate on my idea. We already knew each other since I had previously published in Duda's magazine. At his request, I prepared a memo that better outlined the concept of the Museum. The memo was passed on to Tusk just before his visit to Berlin in December 2007, where—it was expected—his German counterparts would raise the issue of the Center Against Expulsions, commonly referred to in Germany as the "Visible Sign" (*Sichtbares Zeichen*). I wrote,

> The wartime fate of Poles and other Central and Eastern European nations is little known in the West, outside a small group of experts on the topic. The creation of the "Visible Sign" can only solidify this difficult situation by concentrating German and European public scrutiny on the suffering of Germans at, among others, Polish hands. This damage cannot be completely eliminated even by Europeanizing the "Visible Sign," meaning that its narration would present forced expulsions of other nations, including Poles. Previous experience—including exhibitions and publications in Germany—shows that even with such a broad approach, the real causes of events, such as the Second World War and German crimes, would be marginalized. Instead, the resettlement would be interpreted as the result of recurring attempts in Europe from the beginning of twentieth century (or from the Balkan wars of 1912–1913) to create homogeneous nation-states cleared of ethnic minorities. Since it is probably impossible to avoid creation of the "Visible Sign" in Berlin, in one form or another, it is therefore necessary for the Polish side to put forward our own initiative that would depict in this "conflict over public memory" our own concepts, which would go beyond the negation of German projects. One such initiative is the creation of the Museum of the Second World War.

I also pointed out that the implementation of the Museum of the Second World War project would have a great symbolic meaning:

- It would prevent German and European public memory from being dominated by the issue of the expulsions.
- It would show the real reasons behind the expulsions, such as war, occupation, the crimes of the totalitarian regimes of the Nazis and the Soviet Union, martyrdom, and the resistance of Poland and other nations.
- It would expose Western public opinion to Polish historical sensibility and the sensibilities of other Central and Eastern European nations such as the Baltic states, which would certainly join the project. It would be especially valuable to show Western Euro-

peans that, to their European Union's "younger brothers," not only Nazism but also communism was a source of evil.[2]

During his visit to Berlin, Donald Tusk presented Chancellor Angela Merkel with the idea of a museum as an alternative to the museum about expellees. The prime minister also spoke about the Museum of the Second World War in an interview with the *Frankfurter Allgemeine Zeitung*.[3] Still, this did not convince the Germans to withdraw their plans for the "Visible Sign."[4]

I perceived that this rapid adoption of and publicity for my idea expressed an unambiguous desire to resolve the Polish-German impasse on history, but I also saw it as a sign that the prime minister, a graduate in history from the University of Gdańsk, was looking for new ideas at the very beginning of his term. My idea came at an opportune moment when historians were at the helm. Duda was also a historian and Tusk's close friend since the 1980s, when they edited the then dissident *Przegląd Polityczny*. That was why this short newspaper piece that I wrote in November 2007 soon turned into a real project to create a great historical museum. To me, this was quite surprising. Historians publish so many articles, some in journals with large circulations, but most of this writing does not bring about any tangible results. Yet the outcome would be quite different this time.

Over the next few months, at the request of Wojciech Duda and Tomasz Arabski, chief of the Office of the Prime Minister of Poland, I elaborated on the idea of the Museum, including plans for the main, permanent exhibition and the next steps necessary to move the project forward. Donald Tusk decided that the Museum should be located in Gdańsk, the city in which he had been born and spent most of his life. Historical and symbolic arguments supported this choice. First, this is the site of Westerplatte, where—according to the Polish interpretation—the first shots of the Second World War were fired. Second, Gdańsk was also the symbol of the liberation of Poland from communism; it was here that the shipyard strike, which led to the rise of the Solidarity union, took place in August 1980. The newly created European Solidarity Center in Gdańsk was to tell this latter story. Together the two institutions would become a diptych of the history of Poland and Europe in the twentieth century.

---

2 *Muzeum Drugiej Wojny Światowej w Polsce jako przeciwwaga wobec "Widocznego Znaku" w Berlinie*. Private archive of the author.
3 F. A. Z., Gespräch mit Donald Tusk, *Die Geschichte ist wieder Ballast*, December 10, 2007.
4 Jędrzej Bielecki, "Propozycja Tuska dla Merkel," *Dziennik*, December 10, 2007; Jędrzej Bielecki, "Tusk a Merkel o Gdańsku," *Dziennik*, December 11, 2007.

At first, I favored the construction of the Museum of the Second World War in Warsaw, but as I started to get to know Gdańsk, I agreed that the Museum would fit into the city's historical landscape and that it would benefit from the tourist boom in the area. Besides, there were several history museums in Warsaw already, most famously, the Warsaw Rising Museum. Others were being established: the POLIN Museum of the History of Polish Jews, the Polish History Museum, and the Museum of Józef Piłsudski in nearby Sulejówek. Under these circumstances, Gdańsk was indeed a good choice. Tusk also decided that the Museum would be entirely financed by Polish funding and would not be created in a partnership with international institutions such as the European Parliament and the Council of Europe. I did suggest a partnership model in my memo dated December 2007, seeing it as a way not only to ensure that the Museum would have a broader impact on public opinion in Europe but also to secure necessary funding, which may not have been adequate in the solely Polish funding plan. Later on, when the attacks on the Museum of the Second World War began, when we were accused of "cosmopolitanism," "pseudo-universalism," and "serving the interests of Berlin and Brussels," Tusk's decision proved indeed to be the right one. Still, it did not protect us from the allegations of dishonor and disrespect that would trail us in the years to come.

Soon, people around the prime minister proposed that I should undertake the task of creating the Museum. When I had first toyed with the idea, I did not consider for a moment that I would become the Museum's director, committing to a giant task that involved not only creating exhibitions but also carrying out construction and taking on the responsibilities for all organizational, legal, and financial matters. I did have previous experience in establishing a large institution, the research and education division of the Institute of National Remembrance (IPN)[5]. It was a story somewhat similar to that of the Museum. In 2000, I became the director and first employee of the IPN Public Education Office. I defined its operations, hired employees, and finally directed it for over five years. All this took place in an atmosphere of serious controversy, since the establishment and activities of the Institute were criticized by the spokespersons representing leftist and liberal elements of Polish public opinion. The historical research that I conducted at the IPN on controversial issues of great public interest, such as the documents of the Communist security apparatus and material on

---

5 The Institute of National Remembrance, or Polish Instytut Pamięci Narodowej (IPN), established in 2000, preserves the memories of struggle, victims, and damages suffered by the Polish nation during and after World War II; it conducts research to facilitate prosecution of war and state crimes and compensation of those whose human rights have been violated by Nazis and Communists.

the Jedwabne pogrom, instilled in me an appreciation for public history. Without that experience I would not have considered accepting the offer to create, from the ground up, the Museum.

I did understand that this time the task was even more difficult than my work at the IPN. Still, I could not predict what that would encompass over the next few years. I did not really want to make such a dramatic change in my professional and personal life. I conducted research in two institutions: the Institute of Political Studies at the Polish Academy of Sciences, with which I had been affiliated since its foundation in 1990, and at the Nicolaus Copernicus University, where I enjoyed a very rewarding career teaching didactics to nice, ambitious, and intelligent students. I had published a book on Radio Free Europe, the product of years of research, and I had other ideas. I was also close to being named titular professor,[6] which President Lech Kaczyński ultimately awarded me in 2009. So I did not have any motivation to flip my life upside down. If I thought of change at all, I may have vaguely considered the new department of Polish studies being established at Columbia University, which had approached me to explore my interest in working for the department. An appointment at one of the best universities in the world would certainly open new perspectives, personal and professional, but I was not ready at the time to leave Poland for an extended period.

After a few weeks of consideration and discussions with family and friends, I did agree to accept the government's offer. I concluded that this opportunity—creating from the ground up a major history museum that I had conceived—was absolutely unique and would never come again. I had to try it, despite all the uncertainties and the fact that it would completely transform my own life. I took the position of prime minister's special commissioner for the Museum of the Second World War. I was appointed on September 1, 2008, a symbolic date for the institution tasked with telling the story of the war.

Since the Museum was not yet formally created, I could not actually start the preparations, so I agreed to be employed at the Office of the Prime Minister of Poland as advisor to the prime minister. I gained an organizational foothold in Warsaw: an office space and the possibility of holding meetings, as well as the potential to reach out to people who were making decisions on establishing the Museum and on financing its construction. Without these opportunities, the Museum probably would never have materialized; yet my nominal function as adviser to Donald Tusk would eventually be used against me many times. I would be accused of being a politician myself.

---

6 In Polish "Professor Belwederski," Poland's highest academic title.

Museum, but What Museum?

Even before the Museum was formally established in December 2008, one of the first steps was to outline its initial shape and, above all, its exhibitions. This preliminary stage culminated in a study titled *Museum of the Second World War: The Conceptual Brief*, which I prepared together with Dr. Piotr M. Majewski, a historian from the University of Warsaw with whom I had worked previously on the monthly history magazine *Mówią Wieki*.

We stated that while Polish stories should have a special place in the Museum, they would be told in a broader context as part of the history of Europe and the world. We declared at the very beginning of our work that "one of the tools to achieve balance between Polish and 'foreign' issues can be comparative narration. This would allow the Museum to show the similarities and the differences in the character of the war and occupation between Western Europe and Central-Eastern Europe." We also emphasized that the specificity of the war and occupation in Pomerania should be depicted as much as possible, to make the museum part of the local historical landscape, which was itself fascinating and had many universal elements. We also pointed out that we planned to create a museum of war but not a military museum, of which so many already existed in the world. Our attention was centered on the civilian population, the war's main victim. This approach would later give rise to one of the constant lines of attack on our Museum, together with condemnation of our taking into account the experiences of other nations.

We also devoted a lot of attention to the topic of forced resettlement, at the time at the very core of Polish-German disputes.

> The narrative of forced migration should begin with the actions taken by the Third Reich and the USSR from the very beginning of the war and throughout its entire period, and include: displacement of Poles from Pomerania and Greater Poland as far back as 1939; pacification of the Zamość region; displacement of the population of Warsaw after the outbreak and fall of the Warsaw Uprising in 1944; resettlement by the Third Reich of Germans from the Baltic States, the USSR, Romania and South Tyrol; deportations of Poles and the Balts in 1940–1941; as well as the displacement of Germans of the Volga, Crimean Tatars, Ingushians, Karachay, and Kalmyks after 1941. It is in this context that we need to depict the escape of German civilian populations from the approaching Red Army and the displacement of Germans from Poland, Czechoslovakia, Hungary, and Yugoslavia.

Bearing in mind the Polish-German historical disputes, we emphasized that this was one of the most important topics of the exhibitions. It will show that the expulsions of Germans that followed the end of the war were not only an attempt to create an ethnically unified nation-state—as the Union of Expellees would have us believe—but most of all a continuation of the forced migration that was started on an unprecedented scale by the Third Reich and USSR. However, in this exhibition area, it is important to show that even during the most tragic episodes of the German civilian population's attempts to flee the Red Army, the Nazi crimes continued: the death marches from Auschwitz and other concentration camps or the massacre on January 31, 1945 (the day after the *Wilhelm Gustloff* was sunk) of prisoners from a concentration camp in Palmniki [Palmnicken] near the town of Piława [Peilau] on the coast of the Baltic Sea in eastern Prussia, where several thousand people were killed or died during the "evacuation."

We also planned to show in the exhibitions events that were very important to Poland and other Central and Eastern European countries but almost unknown in Western Europe—such as the crimes conducted by Germans against the Polish civilian population in September 1939 and crimes against Poles conducted by Ukrainian nationalists in Volhynia and Eastern Galicia in 1943 to 1944—and also the fact that the end of the war had such a different meaning for various nations. The Museum was to show that 1945 brought freedom to Western Europe, but for Poles liberation from the German occupation was the beginning of a new enslavement: subservience to Moscow and Communist dictatorship. This was only lifted in 1989, when the book was finally closed on the consequences of the war for the nations that found themselves in the Soviet Bloc.

When I read the *Conceptual Brief* today, after the Museum opened, I am struck by the consistency in our many years of work on the exhibitions. On one hand, the Museum's final shape really reflects the most important notions put forward from the very beginning of the whole undertaking. Even some specific exhibition ideas were already outlined at that time. For example, the display of an interactive model of the German Enigma encryption machine, whose code had been broken by Polish mathematicians even before the war, permitted visitors to write their own messages. And the *Gustloff* and the bell from the wreck of this ship, as the symbol of this tragedy, were used to show the suffering of the German civilian population. The display is complemented in the same room by the story of German crimes committed in the last weeks of the war. These crimes are exemplified by the death march from Stutthof and by the Palmniki massacre, an act perpetrated in part by young boys from the *Hitlerjugend*. All these stories were included in the 2008 *Conceptual Brief*.

However, during the eight years that we worked on the exhibitions, the reality around us changed. Public interest in the expulsions subsided. This resulted partly from the decreased importance of this topic in German politics and consequently in Polish-German relations. The forced expulsions—both during and after the war—are an important part of the Museum, but they do not generate strong emotions and controversies; they were not part of the heated discussions that attended the last phase of the construction and the opening of the Museum.

From the very beginning we presumed that the plan for such an important museum would be subjected to public consultation. We shared the *Conceptual Brief* with most reputable Polish historians of the Second World War and with museum professionals, who all evaluated it in a meeting in Warsaw in October 2008. The meeting was attended by dozens of people. For the most part they were positive, even enthusiastic, about the plan. Professor Tomasz Szarota saw in it the potential for "a wise and courageous undertaking that will show the world the fate of Poles during the Second World War, and on the other hand will show Poles the suffering and martyrdom of other nations, the existence of a common destiny in the occupied territories, and the existence of an international resistance movement. In this sense, the future Museum of the Second World War may also fulfill the function of a specific antidote for the typical to us combination of megalomania and inferiority complex."

The *Conceptual Brief* was then published in the *Przegląd Polityczny*.[7] It was also available on the Museum website. Therefore, the process of developing the Museum was very transparent; it was first the object of a debate among professionals and then presented to the general public, an approach that was not at all a standard process in the development of other history museums in Poland.

---

7 Paweł Machcewicz and Piotr M. Majewski, "Muzeum II Wojny Światowej—Zarys Koncepcji Programowej," *Przegląd Polityczny* 91/92 (2008): 46–51; see also in the same issue a discussion among historians, "Wokół idei Muzeum II Wojny Światowej. Zapis dyskusji," 52–62, and a text about the concept of the Museum of the Second World War by a renowned historian and philosopher, Professor Krzysztof Pomian, 62–65.

"Disintegration of the Polish Nation"

The intensity and ferocity of the discussion around the concept of the Museum, which erupted in the autumn of 2008, surprised me and my colleagues. It foreshadowed the ruthless war that the Law and Justice Party declared against the Museum after taking power in Poland in 2015, although, at that time, I could not have predicted it. Even then, the key actors were the same people who would play principal roles in destroying the Museum eight years later: Law and Justice Party leader Jarosław Kaczyński, right-wing historian and politician Jan Żaryn, and journalist Piotr Semka. We were obviously aware of the fact that history has long been a source of great emotion and controversy in Poland, and some of us—those with work experience at the Institute of National Remembrance—had experienced it firsthand. The first act in the war for the Museum lasted several months and, in my opinion, was one of the most important public disputes over the history of Poland in those years. It was a good illustration of the different historical sensitivities about and visions for dealing with the past, which had been clashing with each other for a long time. The debate began with criticism of the Museum's concept in the *Rzeczpospolita*, a daily newspaper with a clear right-wing bias. One of its journalists, Piotr Semka, accused the authors of the Museum's *Conceptual Brief* of embracing a transnational, universalist perspective on the war, which would lead to the blurring of the role Poland had played in it. In Semka's opinion,

> The exhibition would focus on the scourges that plagued societies: displacements, ethnic cleansing, bombing, and massacres. Such a logic of history puts in the foreground the shared fate of societies. Unfortunately, the consequence is to shift to the background a history of the Second World War perceived as a conflict of nations, as the struggles of Nazi Germany and its allies against countries that have fallen victim to their aggression or that have challenged them. The authors of the project emphasize that they are not interested in building a museum of the martyrdom of the Polish nation or the glory of Polish arms.

Semka also warned that "presenting the Second World War as the anonymous suffering of Europeans benefits the Germans and the nations that collaborated with the Third Reich. By emphasizing the importance of individual suffering, the fact that there were nations who took up the fight with Germany, but also those who surrendered to Hitler, loses significance. Such an idea in a country where veterans of September 1939, the Battle of Britain, and the Battle of Berlin still live appears as a bizarre arrogance."[8]

---

8 Piotr Semka, "Dziwaczny pomysł na Muzeum II Wojny Światowej," *Rzeczpospolita*, October

A few days later, *Rzeczpospolita* published another attack on the Museum. Semka and Cezary Gmyz protested against the universalization and "Europeanization" of the planned exhibitions, noting what the effects of such an approach might be:

> The authors of the Gdańsk museum project argue that their concept is triggered by the desire to reach the widest possible audience. Therefore, they propose to give the exhibition a universal character. It is difficult to agree with this notion. Nobody would have dared to propose to Israelis to universalize Yad Vashem. What happens when certain problems are Europeanized can be seen in Erika Steinbach's exhibition in Berlin. German media reported that it did not pay homage only to German expellees—which was true. However, few people noticed that projecting the subject of displacements onto the European background included the Germans in the victims' club.[9]

In their piece, Semka and Gmyz contrasted the project of the Museum of the Second World War with the Warsaw Rising Museum, created in Warsaw in 2004 by Lech Kaczyński, which among conservative Poles was treated as a model of Polish historical politics and museology. This juxtaposition would be repeated in most critiques formulated against the Museum of the Second World War. Our project was also compared to another historical and museum venture, the House of European History, being established in Brussels by the European Parliament. The House was supposed to show the common history of many nations and also met with opposition from the Polish Right. In a critique of the House, our museum concept, and a French-German history textbook, historian of ideas Dariusz Gawin wrote in *Rzeczpospolita* that all three tried "to mold the European memory top-down, which has little to do with the real public memory of individual European nations."[10] His position is even more important and interesting because he was a cofounder and deputy director of the Warsaw Rising Museum, as well as one of the intellectual creators of the concept of conservative historical politics, which had become an important element of the Law and Justice Party policy when it ruled Poland for the first time from 2005 to 2007.[11]

28, 2008. See also a response from Paweł Machcewicz and Piotr M. Majewski, "Jak opowiadać polską historię," *Rzeczpospolita*, October 30, 2008.
9 Piotr Semka and Cezary Gmyz, "Przypominajmy światu Polską historię," *Rzeczpospolita*, November 3, 2008.
10 "Oczekujemy szacunku dla naszej historii," interview with Dariusz Gawin, *Rzeczpospolita*, December 3, 2008.
11 Gawin was a coauthor of a manifesto of conservative intellectuals, in which the state approach to history post-1989 was criticized and in which a new approach was proposed: a more active role of government institutions in using selected historical narrations to reinforce

Gawin pondered the meaning of our intention not to build a museum of uniquely Polish martyrdom. He thought that this idea was "based on the conviction that at all costs myths should be deconstructed, stereotypes refuted, and that this is one of the basic tasks of the Polish intellectual." In this way, Gawin was situating the Museum project in a long-standing current of Polish intellectualism, one characteristic of which was to contest national myths that cocreated the dominant model of traditional patriotism. In the Museum's concept he also saw cryptopacifism and perhaps even a kind of European "political correctness" that avoided glorifying wars and military virtues. "The removal of military issues to the background results," this conservative intellectual speculated, "from the assumption that one should not toy with these types of emotions, because it would lead to accusations of teaching hatred, nationalism, etc."

This was how Gawin interpreted the announcement that, instead of focusing on military issues, the Museum of the Second World War would show the war primarily through the prism of the civilian population and as a political and ideological conflict. Gawin also expressed concern that a large group of Second World War researchers from other countries would be invited to sit on the Museum's Advisory Committee. "If it is true that Polish historians will have only half the votes in the committee," he told the *Rzeczpospolita* daily, "and the other half will go to historians from other countries, I think this confirms the concerns of those who criticize the concept of universalizing the message."

Soon the accusations went even further. Jan Żaryn, at that time my successor as director of the Public Education Office of the Institute of National Remembrance, accused the Museum of subscribing "to ever more popular concepts of building a common European identity at the expense of national identity. Such concepts are a kind of social engineering." After this relatively neutral statement, much more serious accusations followed respecting the choice of Gdańsk as the seat of the Museum:

> I am more and more disturbed by the actions of some representatives of the Tricity [Gdańsk, Gdynia, and Sopot area] community. . . . I have the impression that there is a living memory of the Free City,[12] especially in Gdańsk: a place in a sense extraterritorial, spiritually con-

---

national pride and societal solidarity. See Robert Kostro and Tomasz Merta, eds., *Pamięć i odpowiedzialność* (Kraków: OMP, 2005).

12 The Free City of Danzig (Polish: Wolne Miasto Gdańsk; German: Freie Stadt Danzig) was an autonomous city-state under the tutelage of the League of Nations that existed between 1920 and 1939, consisting of the Baltic seaport of Danzig (now Gdańsk) and nearly two hundred towns and villages in the surrounding areas. It was created in accordance with the terms of the Treaty of Versailles after the end of World War I. Its population was predominantly German, but it had an important Polish minority.

necting, not separating, Europe. Gdańsk's authorities are trying to build a vision of the future of this city as a city with a new, European tradition, the foundation of which is being built before our eyes. . . . Is this the goal of the Museum of the Second World War? And is the Polish identity that is still alive in many families suitable for any ideological use? The project of the Museum of the Second World War in fact suggests the will to build such a new European identity.[13]

There it is: social engineering, which is to build a European identity at the expense of the Polish and national identity. And for this purpose Gdańsk was chosen, where due to the tradition of the interwar Free City, the ground would be particularly fertile for such operations. From the warnings formulated by Żaryn, it was only a small step to the assessment of another historian from the Institute of National Remembrance (I had hired both of them), who saw in the concept of the Museum of the Second World War, juxtaposed with the House of European History, the manifestation of Brussels's imperialism toward new members of the European Union. Bogusław Kopka wrote in the *Rzeczpospolita* daily, "Four years after Poland's accession to the EU we continue to be treated by the heirs of Charles the Great's Kingdom of the Franks (France and Germany) as petitioners, dependent on the goodwill of the German-French covenant. I must admit that the role of Nipper the dog (known to fans of vinyl music from the EMI record label) who listens to the monotonous voice coming from Brussels is not one I like."[14]

Readers also joined the discussion. *Rzeczpospolita* published a letter to the editor from Andrzej Rzewnicki from the city of Nadarzyn, who wrote about the Museum of the Second World War, "This concept must be fought. I believe that our dramatic fate in this war will melt against the background of European events. . . . This concept is not only erroneous but also dangerous for our future."[15]

The leader of the opposition from the Law and Justice Party, Jarosław Kaczyński (who was prime minister until autumn 2007), was also very interested in the Museum. In his speeches in November 2008, he criticized its concept three times in one week, saying, among other things, that the Museum was a tool to "disintegrate the Polish nation." He spoke most heartily during a parlia-

---

13 "Polska wyjątkowość," interview with Jan Żaryn, *Rzeczpospolita*, October 28, 2008.
14 Bogusław Kopka, "Własnym głosem o naszej historii," *Rzeczpospolita*, December 15, 2008. For those unfamiliar with the allusion used in the article: fox terrier Nipper was featured in an artwork by Francis Barrauda, who in 1898 painted the dog listening to the sound of the Edison phonograph. The painting was titled *His Master's Voice*. For a few decades, the artwork was used as a logo of one of the largest vinyl record producers.
15 Andrzej Rzewnicki, "Jak opowiadać polską historię", *Rzeczpospolita*, November 20, 2008.

mentary debate on the evaluation of the first year of Donald Tusk's government, inserting the Museum project into the wider context of international politics and disputes over history:

> There is the matter of everything that is happening in Germany today, which amounts to a redefinition of the moral, and thus also the political, sense of the Second World War. There is a problem, ladies and gentlemen: the Museum of the Second World War. What is it supposed to talk about? About Polish martyrdom [*Applause*] or about the harm done to Germans? Well, ladies and gentlemen, it should talk about Polish martyrdom [*Applause*], about the Holocaust, which concerned Poles, because, to do otherwise, to be true, would be to agree that some vague, no-name Nazis carried out the crimes elsewhere, while in Poland there were Polish death factories—as has been recently pointed out. It is in practice an agreement about such a situation and a departure from the policy that we have taken in this matter and a serious injury to Poland [*Applause*].[16]

Kaczyński combined a negative assessment of the foreign policy of Tusk's government with this criticism of the Museum of the Second World War. He accused Tusk of accepting the status of a younger, weaker partner in his relations with Germany, and he called the appointment of Władysław Bartoszewski as the special commissioner of the prime minister for international dialogue, responsible among other things for Polish-German relations, "the policy of an ugly girl."[17] This last metaphor intended to symbolize Poland's vassal-like submissiveness to its western neighbor.

During a parliamentary debate, Prime Minister Donald Tusk rejected the accusations of the leader of the opposition, addressing Jarosław Kaczyński directly:

> As far as the Museum of the Second World War is concerned, we talked about it for a long time with very interesting people of great authority, both political and historical. Perhaps you, sir, do not remember . . . but the initiative to build the Museum was conceived and is intended as a Polish response to attempts to misrepresent the history of the Second World War . . . [*Voice from the audience: Ooo!*] You have no right to talk about Władysław Bartoszewski, or about Prof. Machcewicz the way you do, or other people who are involved in this project politically and content-wise. You have no right to accuse these people of wanting to act for another state, against Polish memory and Polish history, because it is exactly the opposite [*Applause*]. Nobody gave you a monopoly on patriotism and truth.[18]

---

16 *The Report of the Prime Minister on the State of the Government's Action Plan, a Year After Its Formation*, 29 Parliament Session, Sejm, November 20, 2008, http://orka2.sejm.gov.pl/Sten oInter6.nsf/0/535759F621D11EB3C1257508004B8E2 A/$file/29_b_ksiazka.pdf.
17 *Report of the Prime Minister*, 171.
18 *Report of the Prime Minister*, 172.

Piotr M. Majewski, coauthor of the *Conceptual Brief*, and I tried to respond to the accusations formulated against us, writing press articles and giving interviews. I even faced Piotr Semka and Jan Żaryn on the TV program of the right-wing journalist Jan Pospieszalski, who would also over the next few years turn out to be an ardent enemy of the Museum. Often, however, it was very difficult to defend ourselves from these attacks, and it seemed somehow weird to find myself explaining that I was not an agent of Berlin and Brussels, a traitor, a "cosmopolitan," and so forth. Through the creation of the Museum, we wanted to introduce Polish history to the mainstream of European and global historical memory to show the world unknown and widely misunderstood aspects of Polish history. To this, the Polish Right responded with accusations of harming national identity and the image of Poland in the world. This departure from the facts and ascription to us of the worst intentions subsequently would dog the work related to the creation of the Museum for years. Some of the accusations were so harsh and absurd that they at times gave the whole dispute an almost surreal character. Considering the final outcome a few years later, the appropriate metaphor would probably be the dark atmosphere of Franz Kafka's novels, especially *The Trial*, whose heroes do not know what they are accused of but are increasingly convinced that no defense can be effective and that everything is going to end badly.

The most important support for the Museum's concept came from some outstanding Polish historians, who contributed to the ongoing public debate. Rafał Wnuk, who at that time was not yet working for the Museum, saw a great advantage not only in comparing the wartime experiences of different nations but also in the intention to show the meaning of broader, more universal phenomena through Polish history.

> The attractiveness of Polish history lies in the fact that it is a history of the borderland, which contains elements that are legible for people from the East and the West. An average Frenchman, an American and an Englishman will find in it both elements that are obvious to them (fighting on the side of the Allies, resistance, the Holocaust) and new, incomprehensible elements (Soviet occupation, perceptions of the conferences in Yalta and Potsdam as defeat rather than victory). . . . In the proposal of Paweł Machcewicz and Piotr Majewski, I find the understanding that this unique character of Polish experience can be used to create a universal message, without losing the national specificity. A museum focusing exclusively on Polish martyrdom or resistance would only address a passive visitor, while the presented outline of the concept proposes to create a museum that is a space for discussion, a place where the Polish experience of occupation would become like the "meter

from Sevres,"[19] a reference point for other countries that participated in the Second World War.[20]

Wnuk noted that the Museum's *Conceptual Brief* provided an opportunity to complement a view of the war dominant in Europe with important yet lesser known threads.

> In the collective memory of Western societies, the Second World War is seen as a period of resistance ending in victory—the same image is cultivated in the memory of Russians, Belarussians, and eastern Ukrainians. The memory of the citizens of Central Europe is different. The cruelty of the occupiers and the heroic resistance is crowned not with victory but enslavement. The way of thinking proposed by the authors of the concept of the Museum of the Second World War is an attempt to present an unnoticed Central European experience as equal to the experience of the citizens of the East and the West.

Marcin Kula, a historian and sociologist at the University of Warsaw, accused the critics of the *Conceptual Brief* of wanting to practice "tribal history."

> The critics of the project would essentially want Poland to build a monument to its own sacrifice and its own glory. . . . [M]onuments to victims and heroes should be built. . . . Reflections on history should not be limited to monuments. Museums are part of the world of knowledge. Knowledge should be transnational. It is the problematic history that allows us to overcome the barriers of national historiographies. . . . The more reflective and less commemorative and blowing of its own horn Polish historiography is, the more chance of that we will have. Everything else is, I'm afraid, waving a saber. Such a show has its own provincial charm, but the world does not appreciate it and will not appreciate it.[21]

Another historian from the University of Warsaw, Jerzy Kochanowski, defended the necessity of depicting Polish experiences in the context of the history of other nations. He wrote in *Gazeta Wyborcza*, "The global conflict of 1939–1945 was a complicated game, happening on many levels—geographical, social, economic, technological, ideological, and ethical. Each of these should be matched with corresponding Polish experiences, because without them the picture will be incomplete. At the same time, the specificity of the 'Polish war' will be more pronounced on the 'foreign background.'"[22] Kochanowski ended his text with a

---

19 The expression "meter from Sevres" refers to a prototype and an international standard for one meter, preserved in Sevres, France. Metaphorically in Polish, the expression depicts a point of reference against which global and historical events can be measured.
20 Rafał Wnuk, "Atrakcyjna historia pogranicza," *Rzeczpospolita*, November 18, 2008.
21 Marcin Kula, "Bębenek historii plemiennej," *Gazeta Wyborcza*, January 2, 2009.
22 Jerzy Kochanowski, "O jaką wojnę walczyliśmy?," *Gazeta Wyborcza*, February 4, 2009.

powerful declaration showing what level of rancor the dispute over the Museum of the Second World War had reached: "I hope that both the authors of the project and those deciding about its implementation will withstand the siege. The aggressors are numerous, but they have a very outdated weapon. I believe that because of the stories told in this Museum, the anti-Nazi resistance in Germany will not be overlooked any more, and no one will talk about 'Polish' concentration camps anymore."

Grzegorz Motyka from the Polish Academy of Sciences also contributed to the discussion. "The critics of the museum's concept speak a lot about the uniqueness of Polish history, and at the same time they seem not to believe their own words, afraid of comparisons with other countries." This eminent researcher of Polish-Ukrainian relations argued that these fears were completely unjustified, because

> thanks to such an approach, the uniqueness of our history can be better understood. For example, the comparison of the Warsaw Uprising with similar events in Paris, Slovakia or Prague does nothing to diminish the insurgency. On the contrary, it shows that only Warsaw did not receive significant help. After the outbreak of uprisings in Paris (August 1944) and in Prague (May 1945), the anti-Nazi coalition troops immediately rushed to their aid with all available forces. To help the insurgents in Slovakia, the direction of the Soviet offensive was changed. As a result, the Battle of the Dukla Pass cost the Soviets about 80,000 killed and wounded soldiers.[23]

The discussion about the Museum was a dispute not only over the interpretation of history but also about the attitude toward Poland's neighbors, its place in Europe, and general ideas about what the nature of relations between nations and states on our continent should be. In the violent attacks of Polish national-conservative circles on the concept of the Museum of the Second World War, one can see the reaction of "Euroskeptics" who wanted to defend Polish distinctiveness and "individuality," which is why they treated a project intended to incorporate Polish historical experience into a wider story about the fate of Europe and the world as a threat to national identity. It is no coincidence that criticism of the Museum concept was, by many, associated with a negative stance toward the House of European History in Brussels and the creation of international textbooks of history, already published in a French-German context and at the time also being prepared for Polish-German history. Everything that went beyond the strictly national perspective raised objections. The debate over the Museum, from the beginning, concerned matters much wider and more fundamental than just the Second World War. It was part of an intellectual and cultural dispute

---

23 Grzegorz Motyka, "Takiego muzeum potrzebujemy," *Gazeta Wyborcza*, April 8, 2009.

about the understanding of Polish history and the place of our country in relations with neighbors and in the integrated Europe—or more broadly, the vision of the European continent (including the place of nation-states and the understanding of state sovereignty) and the contemporary intellectual, cultural, and political trends affecting it. Despite some changing views and new themes that arose from time to time in the debate around the Museum of the Second World War, it was throughout, in fact, the same dispute.

However, there was also another, less noticeable aspect to the conflict over the Museum. In previous years, the public role of history as an important factor in building a national and civic community was promoted primarily by the conservative right wing. The government of Law and Justice, in power from 2005 to 2007, made "historical policy" one of the basic elements of its political platform; this distinguished the party from liberal and leftist circles, which it accused of indifference to national history and tradition. The Tusk government's major historical project, namely, the Museum of the Second World War, could have diminished this convenient political argument and taken away the nationalist-conservative claim to exclusive interest in history and even in patriotism, which had been appropriated by many right-wing politicians and intellectuals. One could assume that this was why such a broad and long-lasting campaign against the project of the Museum of the Second World War, and in some cases also personally against the authors of its concept, was undertaken from the very beginning. It was, it would seem, an attempt to block this project at its conception, to convince Prime Minister Tusk that the project was too controversial and that its implementation might bring more political losses than gains. Even if this plan did not succeed, the campaign against the Museum would at least discredit it in some eyes. It is no coincidence that one of the previously quoted historians who supported the Museum expressed "hope that both the authors of the project and those deciding about its implementation [would] withstand the siege."

The first act in the disputes over the Museum reached its apogee over a few months, at the end of 2008 and the beginning of 2009, but in the following years the Museum remained an important topic of public debate. Subsequent acts were generally triggered by various activities undertaken by the Museum of the Second World War. In August and September 2009, the press published the results of a large public opinion survey conducted by the Museum to probe the social memory of the Second World War—the first such attempt in over thirty years and overall the first done in a free Poland after 1989.[24] The re-

---

24 The full analysis of the survey can be found in Piotr T. Kwiatkowski et al., *Między codzien-*

search showed, among other results, a kind of regionalization of memory of the war. Respondents from the eastern parts of Poland, whose families experienced the Soviet wartime occupation, as well as the bloody Polish-Ukrainian conflict in 1943–1944, placed these events in the mainstream of their memory. In turn, the inhabitants of Pomerania, Silesia, and Wielkopolska, the Polish territories incorporated into the Reich in 1939, primarily mentioned German terror and displacement, as well as signing on to the German nationality list (mostly as a result of various pressures exerted by the occupiers) and forced service in the Wehrmacht. Compared to Poles in areas of the General Government—the German zone of occupation established after the invasion of Poland—they were less focused on the memory of resistance and conspiracy, as these were simply weaker in territories incorporated into the Reich and subjected to even more brutal terror and Germanization than in the regions with uncontested Polish national identity. This thread was violently attacked by some right-wing columnists and historians who believed that the image of the war emerging from public opinion research was a threat to the heroic vision of that time, preserved in literature and film and based primarily on the experience of the General Government, partisan warfare, and the Warsaw Uprising.[25]

This dispute can be considered a clash between the "self-appointed guardians of the national tradition," who saw in the new interpretations of Polish history a threat to national identity, and those who believed that the past should be subject to constantly revisited reflections and that tradition must be open to various interpretations, sometimes leading to a revision of understandings that had been considered irrefutable. Otherwise, history would become stale and not very interesting to the majority of the contemporary public. Apart from regional diversity, memory research carried out by the Museum of the Second World War produced other, very important findings. It showed that Poles remembered not so much military activities but rather those experiences that affected the civilian population. Respondents, when asked about family oral tradition of the war, mentioned the following events: poverty, hunger, and scarcity (9.3%); bombing and air raids (8.8%); forced evictions and expulsion from home (8.2%); escape, hiding, hiding from danger, and hiding belongings (7.7%); arrival of the Soviet

---

nością a wielką historią. Druga Wojna Światowa w pamięci zbiorowej społeczeństwa polskiego (Gdańsk: Muzeum II Wojny, 2010).

25 For more information, see Andrzej Nowak, "Co naprawdę przechowuje polska pamięć," *Rzeczpospolita*, August 19, 2009; Paweł Lisicki, "Przeklęta polska pamięć," *Rzeczpospolita*, August 22–23, 2009; Jan Żaryn, "Nasza historia jest wyjątkowa," *Rzeczpospolita*, August 26, 2009; also see my response: Paweł Machcewicz, "W okopach pamięci," *Gazeta Wyborcza*, August 28, 2009.

army in 1944–1945 (7.1%); executions and genocide (6%); and forced labor for the Germans (5.5%).

These results fully confirmed our vision for the museum exhibitions, in which from the outset we had planned to devote considerable space to the civilian population and its experiences of terror brought by the occupiers, bombings, hunger, deportation, and forced labor. This was in keeping with the public memory of the war in Poland, which the self-proclaimed defenders of Polish history did not want to see, attacking the Museum, among other things, for the fact that it paid too much attention to civilians and thereby neglected the military threads of national history.

History Museums: Between Collective Imagination and Politics

The politics around the creation and operation of historical museums are obviously not unique to Poland. The United States Holocaust Memorial Museum (USHMM) in Washington, DC, the world's first historical narrative museum, was certainly politically motivated. In 1977, the United States established relations with the Palestine Liberation Organization and decided to sell weapons to Egypt and Saudi Arabia. This met with very strong criticism from Israel and Jewish communities in America and the world. In order to remedy this political damage, President Jimmy Carter set up a committee that began work on commemorating in the United States crimes committed against Jews during the war. This memorial would also be an expression of US support for the State of Israel and a testimony to Israel's birth "from the ashes of the Holocaust."[26] When this initial idea was transformed into more concrete considerations over the shape of the exhibitions, the pressure began from various interest groups and organizations operating in the United States, including the Polish American Congress, to take into account the tragedies of other nations. In the end, the crimes committed by Germans against Poles were included, whereas the First World War Armenian genocide committed by the Turks was not.[27]

This political genesis did not prevent the USHMM from becoming arguably the most famous historical museum in the world, receiving over 40 million visitors since its doors first opened to the public in 1993. Moreover, the USHMM played a pioneering role in the creation of the so-called historical narrative museum, which is built from the beginning on the basis of a uniform scenario in which artifacts, photographs, and installations are part of a predetermined story. It was also a source of inspiration for museums in many other countries, including the House of Terror in Budapest and the Warsaw Rising Museum.

Undoubtedly, the German Historical Museum in Berlin and the House of History of the Federal Republic of Germany in Bonn were also political creations. They were the flagships of the historical policy of Chancellor Helmuth Kohl—who, it should be noted, had a doctorate in history—after he took power in 1982. These were criticized by opposition parties and liberal and left-wing historians, who accused the ruling Christian Democrats of using history to build a new

---

26 Edward T. Linenthal, *Preserving Memory: The Struggle to Create America's Holocaust Museum* (New York: Columbia University Press, 2001), 17–19.
27 Linenthal, *Preserving Memory*, 49–51, 112–113.

national identity.²⁸ Historian Norbert Frei saw in these efforts an attempt to go beyond the dominant self-critical approach to German history, primarily as a response to coming to terms with the National Socialist period, the war, and the Holocaust. Frei noted that the popularity of the term "historical politics" (*Geschichtspolitik*) in Germany also dates to this point.²⁹ Less than twenty years later, the term would be copied in Poland, although, of course, with different connotations.

An example of the far-reaching manipulation of museums for political purposes is the House of Terror (Terror Haza) in Budapest, dedicated to the terror of the Nazis and, moreover, the Communists. The museum, opened in 2002, is located in a building that was first the seat of the Arrow Cross (Hungarian fascists) and from 1945 the Communist security service. The main message of the exhibition is to equate the crimes of the Nazis and Hungarian fascists with those of the Communists and at the same time to show continuity in the perpetrators' attitudes. In fact, only about 20 percent of the exhibition is devoted to the war; the rest covers the postwar period. This is in sharp contradiction to the real historical balance: Hungarian Stalinism was indeed very repressive, but the number of its victims comes nowhere near the half million murdered Hungarian Jews. One of the most spectacular parts of the exhibition is the multimedia presentation in which an Arrow Cross member transforms himself into an officer of the Communist secret police simply by changing his uniform. Both the Arrow Cross men and the Communists are portrayed as agents of foreign forces—Germans and Soviets. Their crimes do not burden the Hungarian nation, which for several decades was under foreign occupation—first, briefly, German, and then, for much longer, Soviet. The exhibition is, as it happens, silent about the fact that the vast majority of Hungarian Jews were deported to Auschwitz in cooperation with the Hungarian authorities, allies of the Third Reich, even before the Arrow Cross took power in October 1944. Hungary is shown as a victim of two totalitarian regimes, and the alliance with the Third Reich, which allowed it to regain some territories lost after the First World War, is virtually absent from the museum's story. The museum was opened in a hurry, still unfinished, as a main event of the parliamentary elections campaign. Prime Minister Viktor

---

28 Manuel Becker, *Geschichtspolitik in der "Berliner Republik." Konzeptionen und Kontroversen* (Wiesbaden: Springer, 2013), 118–120; Edgar Wolfrum, *Geschichtspolitik in der Bundesrepublik. Der Weg zur bundesrepublikanischen Erinnerung 1948–1990* (Darmstadt: Wissenschaftliche Buchgesellschaft, 1999), 336–340. An interesting analysis of historical policy during the Kohl's administration can be found in Michał Łuczewski, *Kapitał moralny. Polityki historyczne w późnej nowoczesności* (Kraków: OMP, 2017).
29 Norbert Frei, "Geschichtspolitik," *Süddeutsche Zeitung*, April 29–30 and May 1, 2017.

Orbán was interested in showing both the co-responsibility of his socialist rivals for the crimes of the Communist era and, of course, his own merits. The exhibition ends with film scenes from Imre Nagy's reburial in June 1989, when Orbán, as a young, radical Fidesz leader, called for the withdrawal of Soviet troops from Hungary, followed by shots of the last Soviet soldiers leaving the country in 1991. This is finally, in the museum's depiction, the end of the occupation. And in the finale we can see a fragment of the film from the opening of the House of Terror, again with Orbán speaking. So as to leave no doubt, a marble block placed before the entrance to the museum informs visitors that it was created thanks to the prime minister, the leader of Fidesz party. Despite the opening of the House of Terror, Fidesz lost the election. The new socialist government treated it as a creation and tool of Orbán's policy and tried to drastically reduce its budget. In response, Maria Schmidt, director of the House of Terror and a close associate of the leader of Fidesz, threatened to hang large-format photographs of Communist activists and security officers, ancestors of contemporary leaders of the Hungarian Socialist Party, in the front yard of the museum. Under this pressure, the government withdrew its plans to cut the museum's budget. After the return of Orbán to power, the House of Terror became the main tool of his historical policy.[30]

Of course, one can find numerous examples not only of politicians' use of museums and historical exhibitions for their particular immediate goals but also of their further interference in the form of influencing the shape of exhibitions or even changing or closing existing exhibitions. At the Atomic Bomb Museum in Nagasaki in 1996, photographs and texts relating to the murder of a quarter million Chinese prisoners of war and civilians by the Japanese army in Nanjing in 1937 were removed. The museum was supposed to show only Japan's tragedy, not its earlier aggression and crimes. In Nagasaki, this was incited by local conservative politicians, but similar attitudes also prevailed in government circles. The prime minister of Japan ordered a review of all exhibitions in all museums that spoke of Japanese crimes. The official justification was to verify that all photographs shown and facts mentioned were authentic; yet the process has become an excuse to remove uncomfortable content. At the Osaka Peace Memorial, one part of the exhibition was closed for several months, and when it reopened, the story of Japanese crimes in Asia had been replaced by more material on the suffering of Japanese civilians as a result of American bombs.[31]

---

30 Paul Williams, *Memorial Museum. The Global Rush to Commemorate Atrocities* (Oxford, UK: Berg Publishers, 2007), 82–84; István Rév, "The Terror of the House," in *(Re)Visualizing National History*, ed. Robin Ostow (Toronto: University of Toronto Press, 2008), 60–72.
31 Williams, *Memorial Museum*, 121–122.

Russian authorities took even more drastic action toward the most important museum depicting Communist terror. In 1995, in the former gulag Perm 36, where the most famous dissidents were sent after the war, a group of local historians and human rights activists opened a museum in which former prisoners served as guides. It presented the history of the entire gulag system in the Soviet Union. This was possible in the liberal climate of Boris Yeltsin's presidency, when the crimes of the Soviet era, especially from the Stalinist period, were relatively openly discussed. The situation began to change when Vladimir Putin became the president of Russia, a period marked by the gradual and increasingly more unambiguous rehabilitation of the Soviet period, including Stalinism. The museum became the subject of increasingly harsh criticism, and its cofounder, a private human rights society called Memorial, was accused of being a "foreign agent" since it accepted subsidies from abroad. In 2014, the local authorities dismissed the management team that had created the museum from scratch, rebuilding the ruined camp and collecting all artifacts.

In place of the existing museum, a new one was established, and a new director, an official from the local Ministry of Culture, was appointed. Does not that sound familiar? In 2015, the museum opened a new exhibition. From then on, it documented not the repression and suffering of prisoners but the contribution of the gulag system and of forced prison labor to victory over the Third Reich and the service to the country of the officers of the security apparatus managing the camp. The exhibition also stressed how well prisoners lived and worked and described the cultural life that allegedly flourished in the camp.[32] It is hard to find a more drastic example of the interference of politicians and officials in a museum, turning it into a caricature of what it was supposed to be.

In Poland, the museum boom, and at the same time the great role of historical museums in the public sphere and in politics, began with the opening of the Warsaw Rising Museum in 2004. It was one of the first museums in Poland built using modern museological approaches. It offered visitors not only information about the past and the opportunity to see important and interesting artifacts but also a colorful story, full of drama and emotion. It used new technologies, such as films, multimedia screens, interactive stands, and the introduction of sound to the exhibition space. Above all, it fulfilled the expectations of a huge number of Poles that the insurgent uprising would finally be commemorated in the form of a museum. Previous attempts to do so, undertaken after 1989, faced various or-

---

32 Elena Bobrova, "Soviet-Era Gulag Museum NGO Perm-36 Announces Closure," *Russia Beyond the Headlines*, March 5, 2015; Mikhail Danilovich, "Revamped Perm-36 Museum Emphasized Gulag's Contribution to Victory," Radio Free Europe/Radio Liberty, July 25, 2015.

ganizational and legal problems related, above all, to the ownership of the land on which the museum was to be created. At the time, there certainly was not sufficient commitment on the part of the authorities in Warsaw. Still unfinished, the museum managed, on the sixtieth anniversary of the eruption of the uprising, to offer a glimpse of the event to the public; a few months later it was opened to visitors. It stirred enormous interest and, for the most part, received positive and even enthusiastic reviews. Despite greater competition with other newer modern museums in Warsaw and other parts of Poland, the museum, opened many years ago, remains very popular, with an annual attendance of over 500,000.

The opening of the Warsaw Rising Museum was widely perceived as the great success of Lech Kaczyński, the mayor of Warsaw and one of the leaders of the newly created Law and Justice Party. According to popular opinion, it contributed to Kaczyński's victory in the presidential election in 2005 and possibly also to the success of his party in the parliamentary elections held the same year. Certainly, Kaczyński himself, the museum's creators, and Law and Justice juxtaposed this success with the failures of all those who previously held power. In his speech at the museum's opening, Lech Kaczyński stated that Polish society had been waiting for it "for far too long." He also pointed out, though indirectly, who was responsible for that fact:

> Why, however, do we open this museum only on the 60th, and not the 50th anniversary of the uprising? . . . The special character of 1989 and the following years meant that in our times there are still powerful forces for which Polish independence, present in Polish hearts and minds, is not a value but a threat. Forces, for which it is better to attack the uprising than to fight for its memory to the world, for the memory of the facts that for each of us are obvious, but which among other nations are unknown or, worse, completely falsified.[33]

For the Law and Justice camp, the Warsaw Rising Museum has become the most important achievement in the field of commemorating history in Poland since 1989, as well as a kind of model—the meter from Sevres—of patriotism, of presenting the past, and of the "musealization" of history. It was no accident that from the very beginning of the discussions about the project of the Museum of the Second World War, opponents contrasted it with the Warsaw Rising Museum. The latter was presented as an example of the positive attitude of a patriotic institution that depicted Polish heroism properly. According to this reasoning, common among the political Right, other emerging museums should imitate

---

[33] Lech Kaczyński, a speech delivered at the Warsaw Raising Museum, July 31, 2004, wPolityce.pl, www.wpolityce.pl/historia.

the Warsaw Rising Museum because simply nothing better and more Polish could be created. There was never any public discussion about the shape of the exhibitions at the Warsaw Rising Museum before its opening. All critical opinions were condemned as a manifestation of a lack of patriotism, even those which, while positive, indicated some deficiencies in the exhibitions, such as the failure to discuss the wisdom of the uprising and its human and material costs and insufficient consideration of the fate of the civilian population. These were often accompanied by a narrative accusing previous governments of an indifference to history, of disregard for national tradition, and in a still further-reaching version, of a "pedagogy of shame" or "pedagogy of infamy" that concentrated on dark, controversial pages of Polish history such as the crime in Jedwabne, where Poles, incited by Germans, killed their Jewish neighbors in 1941. The break with this "pedagogy" was the creation of the Warsaw Rising Museum in 2004.[34] This thinking omitted, however, all the achievements in Polish historiography after 1989, numerous discussions in the media on the subject of recent history, and, finally, the operation of the Institute of National Remembrance. Since 2000 the latter had been dealing with the history of Poland after 1939, which primarily, due to the Institute's statutory mandate, included its dimensions of heroism and martyrdom. Anyway, after the success of the Warsaw Rising Museum, it became clear that historical museums in Poland are an extremely powerful instrument for shaping the social imagination.

There is no ideal world in which politicians would not want to influence the operations of museums attended by millions of visitors. This is quite an obvious statement. After the success of the USHMM in Washington, DC, and many subsequent narrative museums, it became evident how much influence they may exert on the collective imagination, emotions, and understanding of the past (and thus also of the present). This is even more evident because the museum boom of the past few decades has been accompanied by a corresponding decline in, and sometimes disappearance of, readership. Academic historians no longer hold sway in shaping our ideas about the past, a role increasingly filled by museums that are constantly trying to be more attractive in the sphere of communication to draw mass audiences. This creates stronger temptations for politicians. They are usually the ones making decisions about establishing new

---

34 See the statements by the creators of the Warsaw Rising Museum, Jan Ołdakowski, Dariusz Gawin, Paweł Kowal, and Marek Cichocki, published in *Polityka historyczna. Historycy—politycy —prasa. Konferencja pod honorowym patronatem Jana Nowaka-Jeziorańskiego. Pałac Raczyńskich W Warszawie. 15 grudnia 2004* (Warszawa: Muzeum Powstania Warszawskiego, 2005). The master vision of the historical policy of the political right is described in Robert Kostro and Tomasz Merta, eds. *Pamięć i odpowiedzialność* (Kraków: OMP, 2005).

museums, financing them, and often also hiring their personnel. In many cases, they want museums they create to become their monuments, establishing their place in history. They treat museums as very important instruments for disseminating their worldviews and visions, thereby shaping historical awareness and contemporary political attitudes. The only question is what methods they want to use and how far they are willing to go to achieve these goals. The Russian Perm 36 gulag museum is an extreme example, but an almost infinite range of pressure and interference can be equally effective. Another fundamental question is how historians and curators who create and work in museums and are responsible for shaping exhibitions should address such challenges. How to understand noble but general values in specific situations, such as the freedom of research and exhibition activities and the autonomy of history and culture in the world of politics: that is what this book is largely about.

II Making the Museum

The First Months at the Office of the Prime Minister

On September 1, 2008, I became Prime Minister Donald Tusk's advisor and his special commissioner for the Museum of the Second World War. This enabled me to take the first organizational steps, even before the Museum was formally established a few months later by the minister of culture. I invited Piotr M. Majewski and Janusz Marszalec to join me. Piotr worked at the Historical Institute of the University of Warsaw as well as at the historical magazine *Mówią Wieki*. We became acquainted when Majewski submitted his articles to *Mówią Wieki* in the 1990s. At the time, I was an editor in the magazine's twentieth-century history department. Piotr specialized in the history of Czechoslovakia and the diplomacy of the 1930s and 1940s. He had published very highly regarded and award-winning works on Edvard Beneš, Sudeten Germans, and the Munich crisis. I had known Janusz Marszalec, a graduate in history from the Catholic University of Lublin, since 2000, when we created together the Institute of National Remembrance (IPN). In the summer of that year, after being appointed director of the Public Education Office and becoming its first employee, I found a pile of CVs submitted to the Office by historians who were seeking employment. I chose Janusz, who was the best candidate to create a branch of the Office in Gdańsk. His expertise was in the Polish Underground State, a wartime resistance organization; his PhD, enthusiastically received by historians, dealt with the security forces of the Home Army during the Warsaw Uprising.

After a few months, at the turn of 2009, Rafał Wnuk joined our museum trio. Rafał, was a professor of history at the Catholic University of Lublin as well as head of the Public Education Office in the Lublin Institute of National Remembrance. He was an expert—in my opinion the most knowledgeable in Poland—on the underground independence movement in Poland after 1945, and he also specialized in Polish intelligence during the Second World War. He was a member of the Polish-British government commission to examine the wartime activities of Polish intelligence; as part of this task force, he had conducted archival research in London and Washington, DC. Wnuk was also the coordinator and editor in chief of the monumental *Atlas of the Pro-Independence Underground in Poland, 1944–1956*. The latter was a true magnum opus that chronicled all major actions undertaken by various resistance units. However, Wnuk's critical position on the term "cursed soldiers," used as a token of admiration by Law and Justice followers to describe anti-Communist armed postwar opposition, earned him many enemies on the political Right and in the IPN itself. Rafał Wnuk always spoke against the politicization and instrumentalization of history, and as a leading expert on the postwar underground—not only in Poland but also in the Baltic states—he repeatedly spoke out against the political manipulation around "cursed soldiers."

In 2008, Marszalec and Wnuk no longer felt at home at the Institute of National Remembrance, as they did not accept the vision of its president Janusz Kurtyka, who wanted the institution to represent historical politics and to be an ally of the Law and Justice Party. Independent and moderate staff members were marginalized or encouraged to leave. Many of them left willingly, preferring not to be part of the politicization of the Institute. In anticipation of these developments within the IPN, I left in January 2006, just after Kurtyka took over, in order to return to academia and research activities. I approached Marszalec and Wnuk to jointly create the Museum of the Second World War at just the right moment in their professional lives, when they were open to the prospect of change and building something new. We were united by our age group: I was forty-two at the time, and they were a year or two younger. Piotr M. Majewski was a few years younger than us. The three of us had also experienced together the creation of the Institute of National Remembrance from the ground up and had worked there for many years. Without this ability to function in a state institution, with all its regulations, procedures, and rules for spending public money, none of us would seriously have considered partaking in such a crazy undertaking as the creation of a large historical museum from scratch. Working at the Institute of National Remembrance was, however, in some respects also a negative experience. For us, the Institute was a well-organized yet extreme—one could say Byzantine—bureaucracy, with all its directors, especially of regional branches. We wanted to create the Museum of the Second World War as a kind of antithesis of the Institute of National Remembrance: without bureaucracy, with more direct interpersonal relations, and as a relatively small team connected by a common task rather than a large institution. Although initially very slim, the team grew with time, especially as the opening approached, and it became difficult to fully maintain this "antibureaucratic" style.

The meeting space in Warsaw and the organizational foothold, in the form of a one-person secretariat, were very helpful for the project, which from the beginning was conceived not only as a local enterprise in Gdańsk but also as a nationwide endeavor supported by a network of extensive international contacts. I also knew that the road from the prime minister's initial announcement of the decision to create the Museum to its actual opening would be long and bumpy. If it was to be more than a "virtual museum," we needed to overcome many bureaucratic obstacles and secure substantial funding—not an easy task when the government was consistently attempting to reduce the budget deficit. Being attached to the Prime Minister's Office and my having the title of special commissioner of the prime minister for the Museum of the Second World War gave us a chance at more effective operation.

Yet I would pay the price for that; Law and Justice would persistently allege that I became a politician. I do not consider these allegations to be justified, because I did not deal with any issues that would go beyond the Museum's affairs and other topics related to history. Notably, the same people who raised allegations against me were not bothered by the fact that the director of the Warsaw Rising Museum, Jan Ołdakowski, was a Law and Justice MP and, unlike me, was involved in real parliamentary and party politics. However, as these allegations keep surfacing, I will provide an account of my work within the Prime Minister's Office; after all, it was mainly concerned with the very beginning of the Museum, before the Gdańsk team took over.

In the first weeks, the most important issue was to decide where exactly the Museum would be located and how that would affect another planned project, the establishment of the Westerplatte Museum. The latter had its roots in the first Law and Justice government from 2005 to 2007. It was promoted by then deputy minister of culture Jarosław Sellin, originally from the Tricity area (Gdańsk, Gdynia, and Sopot), which was probably one reason he so fiercely opposed the Museum of the Second World War. From 2005 to 2007, the project to create a new museum at Westerplatte remained in the conceptual stage. Paradoxically, the Westerplatte plans were taken up by a politician of the Civic Platform Party in Gdańsk, Sławomir Nowak, one of the closest and most influential collaborators of Prime Minister Tusk. Nowak and a group of his Gdańsk associates, young activists of the Platform, apparently wanted to repeat the success of the Warsaw Rising Museum and to build their public careers on that project.

Nowak expanded Sellin's idea. A kind of "historical park" was to be built at Westerplatte, using both the preserved objects of the Military Transit Depot, a pre–Second World War Polish military site in the free city of Gdańsk, and historical reproductions. It was to be supplemented with a new marina and other facilities intended to attract Gdańsk residents and tourists to Westerplatte. One of Nowak's close relatives, a Gdańsk Civic Platform councilor and the former head of his parliamentary office, was to become the director of the Westerplatte Museum. The relative was not a historian and admitted that he was not interested in history; this was supposed to be an advantage, allowing him to resolve disputes between professionals impartially. The public did not receive this proposal well. There were accusations that this was a clear example of political nepotism on the part of the ruling party. By the way, I thought that the idea of reconstructing the nonexistent objects of the Military Transit Depot was completely groundless, because it would lead to the creation of a kind of historic Disneyland. The true merit of Westerplatte was its authenticity, preserved guardhouses, and ruins of the barracks, which ought to be protected.

The establishment of the Westerplatte Museum was announced on September 1, 2008, the same day that I became the special commissioner of the prime minister for the Museum of the Second World War. This caused some confusion, which Tusk resolved quickly. He invited me to a meeting in his office (I met him personally for the first time then), which was also attended by the head of the Prime Minister's Office, Tomasz Arabski, and by Sławomir Nowak, Wojciech Duda, and Grzegorz Fortuna. All the participants at the meeting but me were from the Tricity; Duda, Fortuna, and Tusk had published a photographic album titled *There Once Was Gdańsk*, which explored the city's past and was immensely popular. Tusk asked a question about the sense of creating two historical museums dealing with the war in one city. He stated that the Polish state could not afford to finance two such large projects simultaneously and asked the invitees to express their opinions. Apart from Nowak, everyone present strongly indicated that Poland needed a museum that showed the entire experience of the war rather than a seven-day defense of the Military Transit Depot. The nation needed an institution that would influence European debates about history, which was also important in the context of plans to create a museum of expulsions in Germany. There was also the question of where to build the Museum of the Second World War. Two locations were considered: on the Westerplatte itself and on Wałowa Street, right next to the Main Town, where the Wiadrownia district (Eimermacherhof) was completely destroyed in 1945. Since the 1970s, a square and a bus depot had stood there, and now the Gdańsk mayor, Paweł Adamowicz, was ready to donate this site for the construction of the Museum. Westerplatte was rejected due to its considerable distance from the city center, which right away would limit the number of visitors. The plot at Wałowa Street seemed like the perfect choice. It was a few minutes' walk from two tourist destinations: a busy pedestrian promenade on the historic Motława River Embankment (Długie Pobrzeże) and the famous Gdańsk Crane. The Polish Post Office (Poczta Polska), which had symbolic meaning as the site of one of the first acts of the war and of the Polish resistance, was also nearby. By the conclusion of the meeting, the prime minister had decided that the Museum of the Second World War, not the Westerplatte Museum, would be built, and its headquarters would be at Wałowa Street.

As a consequence of Tusk's decision, the Westerplatte Museum, which had existed only on paper since September 1, 2008, was transformed into the Museum of the Second World War, which was formally established at the turn of December of that year. With that act, the Civic Platform's government had resolved the problem of two museums in Gdańsk. Yet the case, which seemed to be closed, would later be used by Minister of Culture Piotr Gliński and his deputy Jarosław Sellin to attack the Museum of the Second World War, something which I of course did not anticipate at the time. Another piece of "shrapnel" from the prime minister's decision was that Sławomir Nowak subsequently tried to force me to employ his protégé as my deputy. I firmly refused, explaining that he did not have any qualifications to become deputy director of a historical museum. In response, Nowak's group in the city council tried to torpedo the donation of the building site to the Museum. Fortunately, Donald Tusk was alerted in time and immediately pacified the head of his political cabinet.

Nowak's project did not materialize; no one was employed in the Westerplatte Museum before it was transformed into the Museum of the Second World War. I mention this because during the attacks on me in 2016, I was accused of causing the people who were employed at the Westerplatte Museum to lose their jobs. This simply did not correspond with the truth. There was also a second allegation, that as an outcome of Tusk's decision in September 2008, the Museum was constructed at Wałowa Street, where, it was claimed, the geology and drainage were exceptionally unfavorable. The site, as Deputy Minister Sellin stubbornly repeated during his later crusade against the Museum, was supposedly a "wetland" or "port" area (two obviously contradictory designations). It was even implied for a long time that at one point, it had been the site of a "water intake for the city of Gdańsk." The latter is completely inconsistent with the facts, which I verified by consulting with experts in the history of the city. Indeed, the Museum was established in the area adjacent to the Motława River and the Radunia Canal; yet the bottom-up pressure of groundwater, not its location, presented the largest engineering challenge. The pressure had to be neutralized by creating a large reinforced concrete structure, a "dry bathtub" in the professional language of engineers, within which the main part of the Museum with its permanent exhibitions would be located. Similar conditions prevailed throughout the Main Town itself, and this had not prevented the construction of St. Mary's Basilica or other buildings, including contemporary ones, directly on the Motława River and the Radunia Canal. To build on more stable ground, one would have had to site the museum at least a few kilometers from the Main and Old Towns. And that would not make sense from the point of view of attracting as many visitors as possible, including foreign tourists. It should be noted that it would be even more challenging to build a museum at Westerplatte; issues with this location also surfaced. On this narrow promontory, surrounded on one side by the port channel and on the other by the sea, the construction of a large museum building would be incomparably more difficult, if even possible.

The meeting with Tusk that I described above was one of my few conversations with him. When I had to resolve important issues having to do with the Museum, first of all regarding the Multi-Year Government Program and its subsequent iterations to address longer-than-expected timelines and rising costs, I went to Wojciech Duda, Tomasz Arabski, and later Jacek Cichocki, who had succeeded the latter as head of the Prime Minister's Office. Duda was from the beginning emotionally invested in the project and understood its importance to Polish history. He often acted as an intermediary between me and Prime Minister Tusk when we needed Tusk's support. Minister of Culture Bogdan Zdrojewski at first approached our endeavor with some reservations, perceiving the Museum of the Second World War as a project imposed on him by the prime minister and those close to him and also as a stress on the resources of the cultural sector that could have been devoted to other projects. After a few years, though, this initial reservation transformed into good cooperation and trust, and I could always count on Zdrojewski's support in difficult situations. As a former mayor of the city of Wrocław, he understood the complexity and enormity of the endeavor to build such a large museum, and he was skeptical from the beginning when I assured him that I would strive to open the Museum in 2014, on the 75th anniversary of the beginning of war.

As an advisor to Tusk, I had one other responsibility beyond the Museum: the Institute of National Remembrance. Under the direction of Janusz Kurtyka, the Institute was causing increasing controversy by clearly associating itself with the vision of contemporary history represented by only one party, Law and Justice. This was expressed, among other ways, in the Institute's attempt to deconstruct the historical role of Lech Wałęsa by exposing an episode from his past concerning his collaboration with the Security Service in the early 1970s. Numerous publications and statements issued by the IPN suggested that the main democratic opposition movement and Solidarity had been infiltrated by spies and informers, which also influenced the negotiations of the Round Table[1] and the democratization of the country. The Institute leaked documents intended to compromise various political figures to right-leaning journalists. After the most recent round of attacks on Wałęsa, Tusk asked me in 2009 to prepare an update of the Act of the Institute of National Remembrance in such a way that would guarantee its apolitical operation and minimize the risk that re-

---

[1] The Round Table negotiations between representatives of the democratic opposition on one side and the government and the Polish United Workers Party on the other took place in Warsaw between February and April 1989. The result was the package of political and economic reforms that led to the democratization of Poland and subsequently to the peaceful dismantling of the Communist regime.

cords of the former Communist security apparatus would be used as political instruments. I knew the Institute inside out; I had cooperated on its creation and directed its research and educational division for five years. I understood very well, then, the direction in which to take its reforms, although this was not easy to translate into an act of law. I prepared the main outline of the reform over a few months with Wojciech Duda and another historian with whom I had worked in the Polish Academy of Science and in the IPN and who had a good knowledge of law.

Our updates to the act essentially facilitated access to the archives by, among other things, making the IPN responsible for creating a complete, public inventory and finding aid and by requiring that requested documents be made available to researchers within seven days from the filing of a request. Both updates were supposed to prevent manipulation of access to contentious documents. In some cases documents had been released in an expedited manner: for example, in 2007 the Law and Justice Party had overnight received documents that were then used against judges of the Constitutional Court.[2] In other cases the documents had not been released for long months and even years.

The most important changes, though, pertained to the way in which the management of the IPN was appointed. Up to that point, the college of the Institute would put forward a candidate, who ultimately would be appointed by the parliament, essentially being elected directly by politicians: MPs, senators, and the president of the Republic of Poland. We suggested a different system in which the most important voice would belong to historians and lawyers. In my opinion the update produced some good, though temporary, effects, contributing to depoliticization of the IPN. The council of the IPN, composed of historians and lawyers, chose as the new president a researcher of contemporary history, Łukasz Kamiński, a man with conservative views who was by no means close to the Civic Platform. During his five-year tenure, he tried to maintain distance and independence from politicians, as well as to carry out rigorous historical research at the Institute. A legislative amendment, introduced by Law and Justice in 2016, eliminated the influence of historians and lawyers over the Institute of National Remembrance and their participation in the election of its president, completely giving the Institute over to the control of the ruling party and the parliamentary majority.

---

[2] See Ewa Siedlecka, "IPN chroni posła Mularczyka," *Gazeta Wyborcza*, August 9, 2007; Michał Engelhardt, "TK kontra Mularczyk. Czy poseł PiS był w ogóle w IPN?," Gazeta.pl, October 23, 2008.

As I have already mentioned, the Prime Minister's Office served as our organizational foothold in Warsaw, especially in the first period before the formal creation of the Museum. Piotr M. Majewski and I were at that time meeting with people who had experience in creating historical narrative museums. I remember a conversation with Jan Ołdakowski and Dariusz Gawin of the Warsaw Rising Museum and with Jerzy Halbersztadt, then director of the Museum of the History of Polish Jews, at that time still under construction but initiated many years earlier. The latter persuaded us that creating a museum required breaking with the habits of academic historians and switching to a completely different way of perceiving the past, space, and objects. Soon we would be convinced that he was absolutely right.

From the very beginning, we assumed that the exhibitions should be created in a dialogue with historians and museologists. The above-mentioned discussion of the Museum's concept served that goal, as did the commissioning of expert reviews by historians and museum professionals about how to present selected problems in the main exhibition of the Museum. We asked the authors of these studies to recommend which important themes within each of their subjects our exhibitions should cover, how to present them, and what artifacts, photographs, and people were especially important and worth presenting. The studies that we received contained both very general observations, valuable for our reflections on the entire Museum and relations between its individual sections, and sometimes very specific advice on what materials to look for and how to acquire interesting artifacts. It was an important stage in conceptualizing the exhibitions at the very beginning of our work.

At this initial stage, we also conducted study tours to various historical narrative museums, especially those we had not seen before; being historians we had already visited as many of them as possible long before the Museum of the Second World War appeared on the horizon. I remember that the Flanders Fields Museum in the Belgian city of Ypres, devoted to the struggles that took place on the nearby battlefields during the First World War but at the same time undertaking a very successful attempt to show the universal experience of the war, made a great impression on us. It certainly was a source of inspiration for us. We visited the Imperial War Museum in London, which was then largely a traditional, fairly old-fashioned military museum with a lot of weapons and uniforms, but we also spent time in the recently opened Imperial War Museum North in Manchester. The latter, in a building designed by Daniel Libeskind, had modern, interactive exhibitions that did not focus on the details of battles and military equipment but tried to show the reality of war for both soldiers and civilians. The Manchester museum has led to the revitalization of a neglected postindustrial part of the city; soon after it was built, more establishments started to appear there, including an art gallery, and the entire district became one of the city's attractions. It was an interesting point of reference for us at the time, for we were beginning to plan a museum in a part of Gdańsk that was virtually undeveloped after 1945, and now we were given the chance to revitalize it for the city.

In France, we saw a very interesting museum of the First World War, Historial de la Grande Guerre, in Péronne, near the battlefields of the Somme. It is, in a sense, a very traditional museum that speaks primarily through the objects displayed at the exhibitions, but at the same time it is tangible proof that the original artifacts can create a fascinating story. What distinguishes the Péronne museum is its balanced and equal representation of perspectives of three states deeply involved in the war: France, Great Britain, and Germany. Of course, this would not be possible in the case of a museum about the Second World War, where the responsibility for aggression and crimes was unambiguous and much more one-sided. In the case of the Great War, however, this approach was historically and morally legitimate, and the exhibition in Péronne was very inspiring and unique.

In turn, in Paris, we saw the latest creation of French museology: the Charles de Gaulle Monument (Mémorial Charles de Gaulle), in a sense the extreme opposite of the Historial de la Grande Guerre. It was a completely multimedia experience, with dozens of screens on which images were constantly changing. Some of them did not work, although the museum had opened only a few months earlier. I had the impression that I was in the newsroom of a big TV station, and I was unable to find one authentic artifact there. To us it was a warning about where excessive fascination with modern media and disregard of original objects could lead.

The trips east were also interesting. In Kiev, we visited the Museum of the Great Patriotic War, erected during the era of the Soviet Union, when it played the role of the most important museum dedicated to the war and focused on military history. After Ukraine gained independence in 1991, it was "Ukrainized." Now it was repurposed to tell the story of the Great Patriotic War in Ukraine and the story of soldiers who came from there. In practice, it was difficult to separate this story from the "omni-Soviet" history. New elements had also been added, beyond the dimensions of the main exhibition: the massacre of Jews carried out by the Germans in Babi Yar near Kiev and the fight for independence by the anti-Soviet Organization of Ukrainian Nationalists and the Ukrainian Insurgent Army. It was an interesting hybrid of old and new approaches that coexisted side by side in the museum but in fact lived as separate parts. After the Russian aggression against Ukraine in 2014, the name was changed to the Museum of the Second World War in Ukraine, and the exhibition about contemporary battles in the Donbass was opened in the museum's hall. When I visited it in 2015, the director proudly showed me the newest exhibit: a bullet-riddled car belonging to the legendary soldiers (nicknamed "Cyborgs") who defended the Donetsk airport. Though not a model for us in terms of its exhibitions, the Kiev museum was living proof of how commemoration of the war changes under the influence of current events and, at the same time, of the difficulty of changing long-established ideas about it.

In Moscow, we visited the monumental Museum of the Great Patriotic War in Poklonna Gora. Its construction began in the 1980s under the Soviet Union, but it was opened in 1995, on the fiftieth anniversary of the end of the war. It celebrated the Soviet and Russian triumph over the Third Reich through a display of captured enemy military flags (the same flags that were thrown in front of Stalin during the famous parade in Red Square in 1945), gigantic dioramas depicting the most important battles of the war, and a Hall of Glory in which were engraved the names of all heroes of the Soviet Union. There was an incredible amount of military equipment, both Soviet and German, often unique. Though this was certainly not the direction in which we wanted to move in creating our Museum, we envied the Moscow museum's access to many valuable artifacts. Our hosts could feel it well.

During the reception hosted for us, the deputy director took my deputy, Piotr M. Majewski, aside. He urged him to consider including a Russian historian recommended by the Museum of the Great Patriotic War, a de facto representative of the Russian authorities, on the Academic Advisory Committee for our Museum. In turn, we could receive from their huge collections all the artifacts we wished for. He argued that it would be very difficult for us to build from scratch a collection that would enable us to create valued exhibitions. Of course, I did not agree to this proposal, but it was an interesting experience for us that showed how closely our work was being observed in Moscow. The hosts knew of the *Conceptual Brief* published on our website and did not hide their disapproval, which paradoxically brought them closer to the Polish political Right.

A little later we arrived in New Orleans, where the largest American museum dedicated to war is located: the National World War II Museum. Though its name is almost identical to that of the museum in Gdańsk, in fact, in concept and shape the two are extremely different. The New Orleans museum deals with the war only from the American point of view. Even in its official materials (for example, various commemorative objects handed out to guests and sold in a museum store), the duration of the war is clearly stamped as 1941 to 1945, from the attack on Pearl Harbor to the surrender of Japan. The exhibitions consist of two blocks: one dedicated to the struggles of American soldiers in the Pacific and the other concerning the European military theater. The experiences of other nations and the fronts on which Americans did not fight practically are not presented at all. About the same amount of space was devoted to Poland as to the Soviet Union: both countries were mentioned only twice. One of the two places where the Soviet Union appeared was on the label for a beautiful, shiny Harley-Davidson motorcycle, with information that the United States had handed over 100,000 such machines to the Soviet Union under Lend-Lease. The civilian population is depicted only in the story of economic mobilization on the home front: for example, the factory work of women replacing men drafted by the army. In a sense, it was a mirror image of the Museum of the Great Patriotic War in Poklonna Gora, and I cannot resist the impression that both museums would meet the expectations of many of our critics, if only the Russian or American perspective could be changed to Polish. Visits to Moscow and New Orleans convinced us that our vision for the Museum was worth implementing, because its exhibitions would be unlike those of any other existent museum dedicated to war.

Assembling the Gdańsk Team

At the beginning of September 2008, I went to see the Wałowa Street site with my guide in Gdańsk, Wojciech Duda. It was striking that although we were a step away from the most touristy and crowded places in the Main Town, one got the impression—as in many other places in Gdańsk—that the war had created an abyss that persists to this day. Here, the buildings ceased, and an empty, paved square occupied by the bus terminal began. It was hard to imagine that soon a museum would be built here (the building's architectural design had not yet been selected, and therefore we did not have the slightest idea what it would look like). Other buildings would appear next to the Museum, and ultimately they would lead to the refilling of the urban substance of this war-ravaged fragment of Gdańsk.

For me it was also the beginning of a long and fascinating process of getting to know Gdańsk and gradually growing into this city. I realized that I was an outsider, which was a disadvantage but also an asset, because the Museum was not to be limited to local history but to present the whole Polish experience of the Second World War against a wide European and global background. My deputy and closest associate, Janusz Marszalec, was intimately acquainted with the place, joined soon afterward by other people who were recruited for the Museum from Gdańsk and the local university. I, in turn, did everything to get to know the city, its people, and its history, step-by-step. After work or on weekends, I walked or biked through Lower Town, Biskupia Górka, Wrzeszcz, the old shipyard areas, and Nowy Port, searching for traces of old Gdańsk but also watching how newer, more contemporary layers overlapped.

For several years, I was a member of the Council of the European Solidarity Center, which gave me insight into another fascinating undertaking and an important chapter in the history of Gdańsk. Soon my wife joined me in these peregrinations. Also a historian, she was working on a book about strikes in the Tricity in August 1980, which she presented against a broad background of the 1970s and even earlier decades. With the zest of the researcher, she now met dozens of people and sought out places she wrote about. By the time eight years had passed, we had come to feel we were deeply rooted in Gdańsk. We have tried to know its history and genius loci, or "local character," more consciously than Warsaw's, where we had been born and raised and where almost everything was familiar or even obvious to us.

After the formal establishment of the Museum by the minister of culture at the turn of December 2008, we were able to create a team in Gdańsk. We had two small rooms, lent to us by the province, through which dozens of candidates passed, usually new or recent graduates who flooded us with their applications and CVs. We chose historians, most often graduates of the University of Gdańsk and the Nicolaus Copernicus University in nearby Toruń, preferably those who had taken courses on the Second World War or at least the history of the twentieth century. Among those who joined the team a little later was a German historian, Daniel Logemann, who spoke excellent Polish. For me, it was very interesting to observe how a German historian fit in perfectly with the Gdańsk team, while fully accepting what could be called the "Polish point of view"—rebutting the main accusation directed against us by Jarosław Kaczyński. When we worked on a joint exhibition on the Second World War as part of a group of European museums under the Liberation Routes Europe network, we argued strenuously with the representatives of the German Allied Museum in Berlin over how to formulate messages about bombing and forced displacement. I remember Daniel, raging and pounding his fist on the table, exclaiming that "we in Poland, by no means, can allow Germans to treat us this way," which became a beloved wisecrack among the Gdańsk team.

The team was made up mostly of young people, historians straight out of university; for many of them, this was their first job. They were enthusiastic about creating the Museum and its main exhibition, understanding that it was an extraordinary and unique professional opportunity. Some tried to work on their PhDs at the same time. Janusz Marszalec, Piotr M. Majewski, Rafał Wnuk, and I tasked ourselves with leading everything in the right direction and at the same time giving the younger staff a chance to develop themselves, something that would benefit the Museum. I was reminded of the atmosphere at the beginning of the creation of the IPN's Public Education Bureau, without the nightmarish bureaucracy that made it a traumatic experience. The salaries at the Museum were not high; we required hard work in an increasingly stressful atmosphere. We encouraged the staff to continue their own research, facilitated trips to archives, and underwrote language courses. Rafał Wnuk, who became at one point the head of the Research Department, conducted a regular seminar at which employees presented their research projects.

Within a few months we moved from the office provided by the province to the very heart of Gdańsk's Main Town, to a tenement house on Długa Street, just next to the Golden Gate, where we rented rooms until we moved to the newly constructed museum building in March 2017. For the first year, ours was a small team consisting of fewer than twenty people. With time, especially after the start of the construction, this number grew to several dozen—at the time of opening the team comprised around sixty people, including a dozen engineers running the construction work. A significant and unexpected barrier in the systematic creation of the team was a wage freeze within the public service introduced in 2010 and lasting until 2015. Decreasing the budget deficit was certainly a top priority for the state, but for an institution just being created, it was a real blow. At one point we practically could not employ anyone new unless someone else left, and staff could not receive raises or bonuses, which over a few years led to a significant reduction in real wages. We had to achieve the task of creating a large museum from the ground up with a very small team that was, in large part, overworked and thinly spread over a wide variety of tasks.

In 2016 the Law and Justice government took advantage of the good financial situation left by its predecessors and lifted the public service's salary freeze. However, at the same time, the Ministry of Culture and National Heritage assigned the Museum of the Second World War such a low budget in 2016 and then again in 2017 that it essentially prevented the hiring of a team that would open the Museum and support its normal functioning. This was an element of the total war conducted against us from the moment that Law and Justice formed the government in 2015 and Piotr Gliński and Jarosław Sellin took over control of the Ministry of Culture. I will write about this later, but it is worth pointing out that, except during the first few months of our existence, we struggled with financial and staff limitations over all subsequent years, hindering our response to the challenges facing the emerging Museum.

In 2009, we had only a dozen or so staff to carry out two critical tasks. The most important was to conduct a competition for the exhibition design. It was impossible to truly start work on the creation of the Museum without selecting a design company that would translate our storyline into the language of exhibition design and physical layout. From the beginning of the exercise we had adopted a strategic position that the Museum must be created in a consistent and planned way—namely, that the architecture of the future building must take into account the character of the exhibitions, not the other way around.

This was the suggestion of Jerzy Halbersztadt, who adopted a similar approach when he created the POLIN Museum of the History of Polish Jews, but also the result of our own observations. We looked at museums, even very well-known institutions enjoying well-deserved popularity, in which the exhibitions were pressed into the frame of a building, either one that already existed, as was the case with the Warsaw Rising Museum, located in an old streetcar depot, or one that was especially erected, as in the case of Daniel Libeskind's iconic Jewish Museum in Berlin. The buildings preceded the exhibitions, which had to adapt to the limitations imposed by the architecture.

We wanted everything to start with the exhibitions so that they would take precedence. From the first weeks of our work at the Prime Minister's Office, we began to prepare a design competition. We had to translate a very general *Conceptual Brief* into specific provisions of both the themes in the exhibitions as well as their spatial form—that is, the most important sections and areas. The competition was international; many leading European design companies were interested in the Polish market, so the information about Poland during the Second World War had to be expressed in such a way that foreign designers understood what we were looking for and what public expectations we wanted to meet.

Piotr M. Majewski coordinated the design competition, along with me, Marszalec, and Wnuk. We succeeded in inviting prominent individuals with diverse backgrounds to the selection board. Famous, Oscar-winning Polish director Andrzej Wajda, graphic artist Andrzej Pągowski, writer Stefan Chwin, and American journalist and author of historical books Andrew Nagorski were, among others, on the Selection Board. In the end, we received ten entries to the competition, including—as it turned out—some from the most renowned design studios operating in Europe. The British were particularly numerous; they were considered the best in this field.

The jury chose a proposal that delighted with its ingenuity and plasticity but also had an intuitive sense of Polish historical and aesthetic sensitivity. Even at this point, there were elements that emulated the atmosphere we wished to create and that would feature in the finished museum a few years later: for example, the presentation of the Polish Underground State with its basement setting and the large uppercase letters forming the words "Terror" and "Resistance" through which the viewer enters the two main parts of the exhibition. This project stood apart from many other, more restrained and "ascetic" proposals, which probably would have appealed less to Polish viewers. The expectations of Polish audiences had been largely shaped by the widely popular Warsaw Rising Museum, which did not shy away from "rich" staging and strong visual and sound effects. Later, I had the opportunity to find out that, for example, many German museologists and historians were surprised by the scale and intensity of the design of our exhibitions. They were used to more traditional exhibitions common in Germany, in which information and photos played the most important role; in general they contained relatively few authentic objects, and reconstructions and artistic installations were rare.

Andrzej Wajda supported the winning design, considering it very "cinematic"; watching visualizations presented as part of the competition work, he recalled his own memories. "In August 1939, I traveled through Gdańsk," the film director recalled. "The whole city was in Nazi flags. I still remember it today. Young people nowadays are faced with a whole lot of images. That is why the contemporary Museum of the Second World War must have such a visual character."[3] It was the beginning of our long cooperation with Andrzej Wajda and his wife, Krystyna Zachwatowicz, an excellent scenographer and stage designer for numerous films and theater performances, who supported the Museum for many years. Andrzej Wajda agreed to tell us on camera about his father, an officer of the Polish army murdered in Katyń[4], and became one of the main narrators of the film about this crime, which was part of the main exhibition.

The Belgian company Tempora, from Brussels, submitted the winning design. Its team was both very international and very Polish. Its main research adviser was Professor Krzysztof Pomian, an outstanding philosopher and historian and one of the leading authorities on European museology, who had left Poland in the early 1970s after the anti-Semitic and antiliberal campaign of 1968. The core of the Tempora team was two Belgian set designers—Christophe Gaeta and Didier Geirnaert—and French museologist Isabelle Benoit, who at the beginning of our collaboration was our main contact. The intellectual brain of Tempora, besides Pomian, was Elie Barnavi, a very interesting figure: a university professor in Tel Aviv, a former Israeli ambassador to France, and a commando in the Israeli army. Once I tried to get some colorful stories from him about this last experience, but he told me that he mainly remembered constant exercises and marches with very heavy equipment. It made me ponder the everyday lives of soldiers, which we wanted to present in the exhibition and which consist of more than fighting.

Tempora was known for museums and exhibitions, which it had created mainly in Belgium and France, including one in Bastogne on the Battle of Bulge, 1944 to 1945, one in Waterloo on the battle of 1815, and in Paris a famous temporary exhibition on the role of religion in the history of humanity. Its largest but until then still not realized project was the Museum of Europe. The part of the project devoted to the post–Second World War period, titled *Europe—It Is Our History*, was presented in Brussels and in Wrocław, receiving excellent reviews. However, when the European Parliament decided to create its own museum, the aforementioned House of European History, Tempora's project unfortunately lost the chance to be implemented. This was a great loss in the opinion of those who saw samples of the team's work. Next to Belgium and France, Poland became the most important market for Tempora. This was a result of having Krzysztof Pomian on its team and also of the fact that, in the last dozen or so

---

3 *Raport z działalności Muzeum II Wojny Światowej za rok 2009* (Gdańsk: Museum of the Second World War, 2010), 23.
4 In April and May 1940, the NKVD, the Soviet secret police, executed 22,000 Polish officers and officials, more than 4,000 of them in the Katyń Forest, near the city of Smolensk.

years, Poland has become the most important museology market in Europe. No other country has established so many new museums, especially historical museums, with such fanfare. Tempora began to deepen its Polish roots, winning contests for the execution of the Sybir Memorial Museum in Białystok and a museum in Poznań devoted to the beginnings of Polish statehood (Gate of Poznań, opened in 2014). Tempora also prepared an exhibition about Cardinal Bolesław Kominek, archbishop of Wrocław, and his role in Polish-German reconciliation, which was shown in several countries.

Apart from the competition for the design of the main exhibition, the most important task at this early stage of our work was to prepare an exhibition at Westerplatte on a very short deadline. On September 1, 2009, ceremonies commemorating the anniversary of the outbreak of the war were to take place with the participation of European and world leaders, the type of event that always attracts a crowd of journalists and TV crews. It was a great opportunity to present both Polish history broadly and, at the same time, our Museum. Within a few months, we prepared and installed at Westerplatte quite a large exhibition, titled *Westerplatte: Resort-Bastion-Symbol*, which showed the entire story of this place, with the events taking place there dating back to the eighteenth century. The main parts of the exhibition were, of course, the Military Transit Depot, the Polish-German conflict over Gdańsk and Pomerania, the defense in September 1939, and Westerplatte after the war as a national symbol present in living public memory but also as one instrumentalized by the Communists. The exhibition was opened on September 1, 2009, by Prime Minister Donald Tusk, who publicly signed the act creating the Museum of the Second World War. Due to the large international celebration, the exhibition immediately had many visitors, and even Piotr Semka, one of our most ardent critics, wrote about it positively in *Rzeczpospolita*.[5] This caused us to somewhat relax our vigilance and gave us hope that a fairly rational dialogue about history and the Museum itself was possible.

I considered the celebration of September 1, 2009, a very successful manifestation of historical politics (as I have mentioned, I do not like this misused term, but it is difficult to escape because it has become so commonplace in Poland). The world's attention was drawn to the fact that the war began with a German attack on Poland, and our wartime experiences were retold—I had an opportunity to give a long interview to the BBC World Service—as was the country's postwar fate. Vladimir Putin, in his speech, defended the signing of the Ribbentrop-Molotov Pact by the Soviet Union, speaking about Moscow's desire to guarantee

---

5 Piotr Semka, "Westerplatte: Kurort-Bastion-Symbol," *Rzeczpospolita*, September 30, 2009.

its own security and recalling that only a year earlier, Western democracies had contributed in Munich to the partition of Czechoslovakia. However, these arguments and the language used still allowed for a rational debate, which would soon become much more difficult due to the evolving situation in Russia. Those from the right wing accusing the Polish democratic state after 1989 of allegedly conducting historical "shame pedagogy" should have listened to the speech given during the celebrations by President Lech Kaczyński, who unexpectedly apologized for Poland's participation in the partition of Czechoslovakia in 1938 and the occupation of the part of Silesia (so-called Zaolzie) that had belonged to that state.[6]

Returning to Westerplatte itself, the permanent exhibition *Resort-Bastion-Symbol* is visited by tens of thousands of people every year. The Museum also created a second permanent exhibition on the site, telling the history of the Military Transit Depot, including its nonextant objects. In addition to these two outdoor exhibitions and extensive educational activities geared especially toward schoolchildren, there is a branch of the Historical Museum of the City of Gdańsk in Guard House No. 1. The exhibition located there was modernized in 2013 with considerable funding and the substantive support of the Museum of the Second World War. After the establishment of a new Westerplatte Museum in December 2015, the Law and Justice minister of culture, Piotr Gliński, and his deputy, Jarosław Sellin, persistently claimed that the site had been neglected for years and that there was nothing but "bushes" there. These accusations were completely untrue. I have had just one opportunity to talk with the director of the Westerplatte Museum who announced that he plans to create something completely new at Westerplatte and to remove both our exhibitions. So far nothing new has been built there. Incidentally, our exhibition *Resort-Bastion-Symbol*, consisting of huge concrete blocks reminiscent of the remains of military objects from the time of the Military Transit Depot, will not be easy to remove. Perhaps the sappers will have to blow it up, which would be an interesting sight to see.

---

6 Speech given on September 1, 2009, at Westerplatte, *Gazeta Wyborcza*, September 2, 2009.

Architecture, Archeology, and Construction

In 2009 the city of Gdańsk handed the site at Wałowa Street over to the Museum, and it was now possible to start preparations for the architectural competition. An architectural project should refer to a specific place: that always creates an important context for the building and its interior. The value of the land was estimated at over PLN 53 million (€12.4 million), and its donation was a great gesture of support for the Museum from the Gdańsk authorities, especially Mayor Paweł Adamowicz. From the very beginning until the dramatic epilogue, he was our ally; without him, it would have been difficult to bring the Museum to life. The land donation agreement would prove the main argument for preserving the very name of the Museum of the Second World War when, in April 2016, Minister of Culture Piotr Gliński unexpectedly announced the intention to formally liquidate our institution, name and all. Right next to the place where the Museum was to be built, there stood the building of the Polish Post Office (Poczta Polska), next to Westerplatte the most important symbol of the Polish resistance in Gdańsk in September 1939. Günter Grass made its defense and the execution of postmen known to millions of readers in *The Tin Drum*. On the other side of the Museum site, one could see the buildings, especially the tall cranes, of the Gdańsk Shipyard, as well as the European Solidarity Center being built on the former shipyard's lands. There are also plans to build a new district there—the Young City. Add to this the picturesque nature of the place—the location on the Motława and Radunia Rivers—and the immediate vicinity of the Main Town of Gdańsk, and it would be difficult to find a better place to build a new museum and attract crowds of visitors. All these considerations influenced the selection of this location.

We invited prominent architects, town planners, artists, museum specialists, and experts in the history of Gdańsk to join the jury for the architectural competition. The goal was, of course, to choose outstanding architecture that would complement the unique space of the Main Town of Gdańsk in a valuable and interesting way and not disturb its historical character. Therefore, detailed information about the exhibition content and Tempora's layout and design were part of the documentation for the architectural competition. Jack Lohman, a leading authority on world museology and at that time director of the Museum of London, oversaw the solutions proposed by architects from the point of view of the functional needs of a modern museum.

The most important voices, of course, were those of the architects assessing the value of the submitted projects, as well as the urban planners evaluating how the plans fit into the existing urban fabric and affected its development.

Daniel Libeskind, a native of Łódź and fluent in Polish, created fascinating museum buildings erected in many countries, including the Jewish Museum Berlin, the Imperial War Museum North in Manchester, the Bundeswehr Museum of Military History in Dresden, and the Royal Ontario Museum in Canada. He was later the main architect of the project to rebuild the World Trade Center space in New York and commemorate the victims of the September 11, 2001, attacks. Hans Stimmann was the main urbanist in Berlin for several years after the unification of Germany and had a decisive influence on what the city is today. Nota bene: he was criticized by many other architects for being too conservative and advocating for restoring the prewar shape of the city. British architect and urban planner George Ferguson was responsible, among other things, for the revitalization of Bristol's waterfront. Polish architects and urban planners brought, in addition to their professional skills, an in-depth knowledge of the local terrain and context. One of them, Professor Wiesław Gruszkowski, ninety years old at the time of the competition, was one of the main authors of the reconstruction of Gdańsk after 1945 and thus a great link between new and old times. A few years later, when the Museum was almost ready, he donated the Polish flag that he had hidden from the Soviets in Lviv after the invasion of the Red Army on September 17, 1939, and brought to Gdańsk after the war.

The letter inviting architects to participate in the competition was issued by Prime Minister Donald Tusk, underlining the importance of this emerging museum to Poland. He also passed on to me his request to choose a project that would guarantee world-class architecture. He believed that since the Polish state would allocate such significant sums for the construction of a museum, something of value should be created. Gdańsk deserved more than the trivial, repetitive solutions that characterize the majority of public buildings, not just in Poland but elsewhere. Over one hundred proposals from more than thirty countries were finally submitted. Many of them were very standard, placing the Museum in large square or rectangular blocks. Some referred to the types of buildings characteristic of the Main Town of Gdańsk—the work that eventually received second place proposed a building imitating the row of nearby tenement houses located by the Motława River. Others referred to the subject of war. Several proposals presented the Museum building as a ship or bunker; one envisioned it as a tank. One could also be read as a concentration camp, with wires and chimneys. Another, presented by Russian architects, placed on the roof of the building a visually dominant forest of crosses. Had this project been chosen, perhaps Metropolitan Archbishop Sławoj Leszek Głódź would have been more sympathetic to our Museum, rather than accusing us, without visiting the exhibitions (which was sadly the rule for our critics), of devoting too little attention to the martyrdom of the Catholic clergy. The designs can be viewed in the architectural competition catalog that we published.[7]

Of all the proposals, the vast majority of jury members from the beginning favored the one that eventually won, authored—as it turned out—by Gdynia's architectural studio Kwadrat. This project placed the most important part of the Museum, along with the permanent exhibition and the temporary exhibition galleries, underground. The Museum's exterior featured an inclined tower forty meters high, built in the shape of a prism and partially made of glass. It was one of the most daring, original, and unconventional proposals among over a hundred rated by jurors. The jurors appreciated both its originality and great functional solutions to the Museum's needs. The authors of the winning project explained that placing the exhibition underground had symbolic significance—visitors descended down into the hell of war and then climbed back to the surface, overcoming death and returning to life. From the glass tower, they could admire the panorama of the city, risen again from the destruction of war. The red color of the Museum's walls and the distant silhouette of its tower referenced the Gothic churches that dominate the historical landscape of Gdańsk. As the jurors wrote in the justification for their verdict, Kwadrat's project met all the conditions to become a sophisticated symbol, as much as Gdańsk's Armory, St. Mary's Church, and the Crane.

My euphoria after the selection of such a wonderful project, enthusiastically accepted by the vast majority of architects, was cooled by people more experienced in this matter. Daniel Libeskind warned me that until construction started, there was no guarantee that the Museum project would materialize, regardless of the enthusiasm that might accompany all earlier stages. He spoke from the experience of a long-term struggle to realize his project of the Jewish Museum Berlin, which did not enjoy the support of the city authorities. Even then, Libeskind could not have foreseen that starting construction, or even bringing it to the final stage, did not necessarily guarantee that his museum would actually open. The most important task was to get big money for the construction of the Museum, which became even more difficult after the global financial crisis began in 2008. The threat that this might affect the Polish economy became real, which further strengthened the finance minister's aversion to disbursing gigantic sums to a venture that could be postponed and perhaps even completely abandoned.

---

[7] Alicja Bittner et al., *Muzeum II Wojny Światowej w Gdańsku. Międzynarodowy konkurs architektoniczny* (Gdańsk: Muzeum II Wojny Światowej w Gdańsku, 2010).

However, Donald Tusk decided that the Museum would ultimately be built. I think that this was a unique decision among all the countries of the European Union, which were struggling with the crisis or fearful of its consequences and therefore limiting budgetary spending on projects that were not a priority for the economy. When confronted with that, culture, not to mention history, is usually in a weak position. Tusk's Multi-Year Government Program was decreed in January 2011; it earmarked PLN 358 million (€83.5 million) for the creation of the Museum, spread over several years. Now it was clear that we would not remain one more "virtual" museum, and soon we could start the construction.

First, however, it was necessary to carry out archeological research on a vast area of 1.7 hectares. The research lasted about a year and brought fascinating results. The work unveiled a network of streets and the outlines of houses existing since the seventeenth century that were completely destroyed in 1945 during the battles for Gdańsk. It was called Wiadrownia (Eimermacherhof) for the guilds of bucket and jug makers who settled there. It was a poor part of the city but very picturesque, located between the Motława and Radunia Rivers and sometimes called, probably with a bit of exaggeration, Gdańsk's Venice. It was an extraordinary experience to walk over the pavement of Wiadrownia's streets, uncovered after all these decades by the archeological excavations, and to compare this somewhat lunar landscape, evoking imagery of Pompeii, with prewar photographs of the district bustling with life. We transferred the pavement to the interior of the Museum and arranged an alley on the floor that crossed the entire main gallery. It accurately mimics the course of the most important street of Wiadrownia, Wielka Street (Grosse Gasse).

Thousands of items were found during the excavations, some of them even from the seventeenth century, which prompted us to prepare an additional permanent exhibition about the history of the site, titled *There Once Was Wiadrownia*. Its last section dealt with the history of the Museum of the Second World War in this place. This story was the first victim of the censorship imposed by the new museum management after I and my colleagues were removed in April 2017. Gone was the multimedia presentation on what was happening in 2016 and 2017, the aspirations of Minister of Culture Piotr Gliński to liquidate the Museum, our defense, and the protests of public opinion.

Construction started in August 2012. First, it was necessary to register and tender the contracts for the first stage of construction. It was the first of numerous tenders that we had to carry out. All of them were complicated, required huge preparations, and took a lot of time, but none of them was successfully contested or annulled, which I think was a great achievement, considering that historians directed the construction of the Museum. In the case of the European Solidarity Center or the Museum of the History of Polish Jews, investments were carried out by specialized urban agencies in Gdańsk and Warsaw with extensive experience and professional staff. Nevertheless, we still had to create a team of lawyers and engineers. Only gradually were we becoming aware of what a monstrous and incredibly difficult task we were undertaking—it was at the time the largest investment in the field of culture in Poland and one of the largest in Europe. It was too late to retreat to the safety of research work.

The most difficult stage of construction was the first, which required the creation of a huge dry excavation or dry bathtub, as engineers called it in their jargon. It was the underground, the most important part of the Museum, where the main gallery, temporary exhibition space, conference and cinema rooms, warehouses, and parking were to be located. Engineers had to neutralize the pressure of the groundwater, which presented a bigger problem than the neighboring rivers, and also deal with clearing more boulders from the bottom of the excavation than predicted based on preconstruction surveys. During this first period, the building looked lunar: the excavation was filled with water, which stabilized its construction while the concrete side walls and base were being reinforced. Barges with specialized equipment floated on our monstrously large swimming pool, and some of the work was carried out by divers.

This part of the construction turned out to be more expensive than planned and also lasted longer. Postponed deadlines, for which we were fiercely attacked by Law and Justice politicians and right-wing commentators, stemmed not only from technical difficulties but also from a conflict with another investment underway on the neighboring plot. A new sewage collector for a large part of Gdańsk was built there with money from the European Union. After technical review, it turned out that both structures are so close to each other that continuing our work before completion of the collector could endanger its stability. Therefore, we had to stop our construction, initially only for three months. It turned out, however, that the collector construction was delayed by an additional six months; hence we lost nine months at the very beginning of our endeavor. These were not the only unexpected events affecting our schedule. In the final phase of works, there was a construction disaster in the vicinity of the Museum. During the construction of Nowa Wałowa Street, the land collapsed under the intersection, which was of critical importance to us for the delivery of building materials. For three months we had to use a detour, which slowed down the pace of work. I drew from these experiences two conclusions: first, that a great deal is being built in Poland, which sometimes leads to unexpected problems, and second, that such a project teaches humility and is burdened with unpredictable risks. Providing the opening date several years in advance, therefore, does not make any sense. Unfortunately, this knowledge came too late, which is quite natural.

It is difficult to describe what it meant to supervise the construction every day. It constantly generated thousands of problems and issues to be resolved. Each of them had, in general, its legal consequences, which had to be carefully considered, because, after all, we dealt with public funds, the use of which was subject to various restrictions. Certainly, it is easier for private investors who are not bound by public procurement or finance laws. When the main exhibition started to materialize, the challenge was to reconcile work on it with the ongoing building construction. Theoretically, everything was foreseen in the projects, but in practice solutions had to be found for many new conflicts that arose. Janusz Marszalec oversaw all this every day. I have a huge and unflagging admiration for him; he was able to take in and understand all the construction and legal complexities, and at the same time he coordinated the entire investment very well. My role was to make key decisions after listening to engineers and lawyers who repeatedly made conflicting recommendations. I realized that, in the end, I would be responsible for everything, so I was forced to venture into sundry technical and construction nuances of which, in my life up to that time, I had been blissfully ignorant.

The increase in costs was a serious problem. On one hand, the dry excavation turned out to be more expensive than expected, and on the other the situation in the construction market had changed from when we first forecasted a preliminary, comprehensive construction budget in 2009 and 2010. After a wave of bankruptcies among companies specializing in highway construction, which had won public contracts due to undervalued bids, the market cost of public tenders went up. In addition, after the most difficult phase of the financial crisis was over, there was a natural increase in wages and the prices of building materials. It affected us, because in the next tender, for completion of the construction works, we received higher bids than the Multi-Year Government Program had planned for. In this situation, assistance from the minister of culture, who guaranteed additional funds in his budget, enabled us to continue the construction. Our funding was eventually increased by PLN 90 million (€21 million). This would not have been possible without the help of Jacek Cichocki, head of the Prime Minister's Office, who, after Donald Tusk left for Brussels at the end of 2014, was our greatest supporter. Ewa Kopacz, who replaced Tusk as prime minister, did not show much interest in our Museum.

In the end, the construction of the Museum cost PLN 443 million (€103.3 million); we did not even use the entire additional sum granted to us in 2015. Construction lasted four years and five months, and we took possession of the building in January 2017. We had made very good time, if one takes into account the gigantic scale of the undertaking, especially when we subtract about a year of downtime due to the conflict with the sewer collector and the disaster at the construction site of Nowa Wałowa Street. As for the costs, we were accused of being the most expensive of all the museums established to that time in Poland. This is true, but it must be remembered that it is also the largest historical museum of all that were constructed. The judgment on how well the taxpayers' money was spent belongs first and foremost to the visitors.

I can also add that the flagship project of Law and Justice, the Polish History Museum, which is being built at the Citadel in Warsaw, is budgeted to cost PLN 757 million (€176 million), which is almost twice as much as the Museum of the Second World War.[8] It is simply impossible to construct a large historical museum with modern world-class exhibitions for a much smaller amount.

---

[8] Uchwała Zmieniająca Uchwałę w Sprawie Ustanowienia Programu Wieloletniego "Budowa Muzeum Historii Polski W Warszawie," Premier.gov.pl, March 20, 2018, www.premier.gov.pl.

Building the Collection and Planning the Exhibitions

The most fascinating part of our work was creating the main exhibition, a task to which we were, as historians, somewhat better suited than overseeing a major construction project. Despite that, we had to significantly change our way of thinking about history, shaped as it was for the vast majority of academic historians by the heavy volumes we had read and sometimes written, by archives, and by the documents found in them. Our thinking had revolved around facts, ideas, and historical processes. Now, we had to begin to understand and present the past in a different way, one understandable and attractive to the general public, which mostly has little knowledge of history. We had to learn by ourselves that the most important thing in the exhibition is the object, the photo, the symbol, and sometimes the reconstruction or dramatization of some place or event. In this way, on the one hand, you can convey the most complex intellectual content, and on the other, you can evoke various emotions in the audience: passion, tenderness, amazement, sadness.

We gained a lot from our cooperation with the museologists and stage designers from Tempora, who initially clashed with our academic rigor and expectation of putting as much content in the exhibition as possible. Among other things, they forced us to be concise in our labels, imposing a maximum number of characters for each type of text, from the introduction to the exhibition, through the main text in individual sections, to the tombstones and captions for individual artifacts. It was probably the most painful process for all the historians on our team. After preparing scenarios of several dozen or even, in some cases, several hundred pages for particular parts of the exhibition, we had to write an introductory text in a few sentences that would be comprehensible and interesting to the average visitor, not just those from Poland.

Here are examples of the very concise texts introducing the two main parts of the exhibition, which were supposed, on one hand, to convey the most important messages within the presented subject and, on the other, to encourage visitors to explore more detailed levels of narrative.

Terror

Coercion as the underpinning of occupation policy

Repression was an inextricable component of occupation in all the subjugated countries; it differed in type and dimensions. Occupied Poland was in a special situation, since already in 1939 both Germans and Soviets began to institute indiscriminate terror against people considered political, racial or class enemies. With time, executions, mass deportations of whole populations, transportation to perform forced labor, incarceration in prisons, ghettos, concentration camps and forced-labor camps, as well as pacifications of entire villages and towns in retaliation for resistance, affected every citizen of nearly all the occupied countries to some degree.

Resistance

Various forms of struggle against the occupying powers

The conquered nations responded to the occupation by offering resistance. It took on various forms in different countries and situations, ranging from listening to banned radio stations and reading underground publications, through patriotic demonstrations, taking part in underground teaching, assistance to people in hiding all the way to sabotage and partisan warfare, and even insurrections. In occupied Poland, the establishment of an underground state, with its own administration, law courts, educational institutions and army, was unique. The price paid for resistance were the bloody repressions, with which the Germans punished entire communities on the principle of collective responsibility.[9]

Initially, such summarization seemed to us an incredible reduction and trivialization of the historical record. However, the labels had their various levels, providing different degrees of detail. If the interested visitor wanted to know more, he or she could go a bit further, as into a labyrinth, and read the artifact captions, some of which were extremely detailed (such as those describing weapons) but often also containing more general information. Finally, there was multimedia, where much more information could be stored than on labels. Even with the utmost brevity, which we found very challenging, the panel text, captions, and tombstones counted about 330 pages of typescript, and there were about 1,200 pages of text in multimedia stations. In total, this amounts to several books, impossible to take in entirely during one visit to the Museum. It is hard to expect that even one visitor would be interested in all the content present in the exhibition. The large exhibition should offer various levels and sets of information that visitors will choose independently, without losing—this was the basic premise—the big idea of the Museum as expressed in the primary wayfinding strategy. Writing and editing the captions indeed took several years. I worked together with Piotr M. Majewski, Rafał Wnuk, and Janusz Marszalec on the final version of the text, chiseling every word to perfection, shortening it as much as possible, all along wondering if it remained understandable for people without historical knowledge.

Yet we soon realized that many fundamental issues were best presented through specific, poignant objects and stories of individuals and families. A

---

9 The English text is a direct quote and the official translation of the Polish exhibition text.

good example is one of our most exceptional artifacts: a handkerchief belonging to Bolesław Wnuk, a rural activist of the Zamość region and a member of Parliament, who was arrested by the Germans in November 1939 and murdered in June of the next year as part of the AB-Aktion.[10] Imprisoned in the Lublin castle, Wnuk learned from the Polish guard that he would be shot the next day. On a handkerchief, later passed on by the guard, he wrote a farewell to his family:

> Dear wife, Niuniusz, Laluniu, Grzesiu, mother, sister, brothers and sisters-in-law, relatives, friends, today I am being shot by the German authorities. I'm dying for my homeland with a smile on my lips, and I am dying innocent. For my blood, may God pay the scoundrel with an eternal curse,
>
> Your Bolek

Additionally, using photographs, we recreated Wnuk's family history, showing his younger brother, Jakub Wnuk, a doctor of pharmacy and a Polish army officer, who was taken prisoner by the Soviets in September 1939 and sent to the camp in Kozielsk. He was shot in April 1940 in Katyń. In the immediate vicinity of the Wnuk brothers, we showed the story of two sisters, daughters of General Józef Dowbor-Muśnicki. Janina Dowbor-Muśnicka Lewandowska was a pilot and paratrooper. Together with a group of military pilots, she was taken prisoner by the Soviets. In April 1940, she was shot by the NKVD in Katyń. The younger sister, Agnieszka Dowbor-Muśnicka, became involved in the work of the underground military organization "Wolves" in the first days of the German occupation. In April 1940, the Gestapo arrested her in Warsaw. She was shot in Palmiry forest, near Warsaw. In this case, the artifact that led to the story of the sisters was a brass eagle from the head of the banner of the First Polish Corps, created and commanded in Russia by their father in 1917. These two family stories, complemented with exceptional objects, were a more expressive and moving testimony of German and Soviet terror than the dozens of pages of information we could have placed in this part of the exhibition.

The Tempora team consisted mainly of foreigners, which turned out to be a great advantage in creating an exhibition that was to be understandable and interesting not only to Poles. The designers and museologists from Belgium were the first audience with which we could test whether our storyline was too esoteric. Often, after discussion with them, we reduced the amount of information, changed the wording, or combined Polish experiences with equivalents from other occupied countries, which provided a better understanding of what was

---

10 AB-Aktion, also called the Extraordinary Pacification Operation, was conducted by the Germans in 1940 to exterminate Polish intellectuals, political and community leaders.

happening in Poland. We quickly accepted the assumption that all the information in the exhibition must be understandable to a visitor who might know nothing about the history of Poland. An imagined tourist from Portugal was our virtual future visitor, who was supposed to restrain our temptation to give too much information or overly complicated explanations. Why from Portugal? It lies on the antipodes of the European continent, it has almost no historical ties with Poland, and it did not participate in the Second World War. Many times, when we delved into detail while working on an exhibition, we were called to order by asking whether what we were preparing would be understandable to our Portuguese visitor.

It should be noted that in Gdańsk, especially after Euro 2012, one constantly saw and heard Spanish tourists, but for a long time, I could not find the Portuguese, and I began to wonder if our choice was too dogmatic. Fortunately, since 2015, for reasons unknown to me, I have finally started to meet numerous Portuguese tourists on the streets of Gdańsk.

This anecdote shows one of the most important challenges facing the creators of the exhibitions. They should be interesting to a Portuguese tourist who knows nothing about Poland, and also to a Polish high school student who already knows something about our past (generally not too much), and also to a historian with deep knowledge. They must then be subject to various levels of interpretation, while their elemental message is clear to everyone.

After the competition for films and multimedia presentations was settled—the jury included well-known Polish documentary film makers—the work on the exhibitions gained another dimension. Within the Museum's scope there were over 250 film and multimedia presentations, totaling about seven hours. Their preparation was a gigantic challenge that had to be met by the winning contractor, NoLabel, which had previously created, among others, the greatly popular underground exhibition in the Market Square in Kraków.

The experts from NoLabel had thousands of discussions with our team of historians about every detail of the material being prepared, including background music, the choice of which was often not straightforward. Frequently, we analyzed one film or presentation several times, sending it back for corrections. In the last dozen years or so, primarily since the opening of the Warsaw Rising Museum in 2004, multimedia has become a kind of fetish in Polish museology. The audience often evaluates the exhibitions through that very prism, considering as modern and attractive only those with numerous multimedia productions that provide the best and most surprising audiovisual "special effects." We considered this direction too kitschy and wanted to create our exhibition in opposition to this trend, placing emphasis on authentic objects. We had the feeling that the wave of fascination for multimedia exhibitions would subside. They cease to be a novelty, especially when almost all of us have access to a virtually unlimited multimedia resource on our computers. Nevertheless, we could not completely ignore the expectations of the audience for whom the Museum was built. We did not decide, however, on the equally radical step taken by the Museum of Warsaw, opened in 2017, which exhibits only artifacts. In a narrative museum presenting such a multistrand event as the Second World War, it would be impossible.

Still, we tried to take a pragmatic approach toward multimedia elements. They were not an end in themselves; we did not want to dazzle the target audience with them, but they remained an important supplement to the exhibition. It was possible to put a lot of information into them, providing specialized content (e.g., for enthusiasts of military technology) for which there was no place in the main visitor path. On the multimedia screens we presented all military campaigns, the technology allowing us to clearly depict directions of attacks, the changing lines of the front, and an enormous number of photos from the battlefields. In some parts of the exhibitions, films and multimedia presentations played a more important role. We wanted to show communism, fascism, and National Socialism through the prism of their ideologies and propaganda, which seduced and pushed millions of people to crime and war: hence the great significance of the imagery and music from the epoch, such as the voice of Mussolini or fragments of films by Hitler's court director Leni Riefenstahl.

The final chord of the exhibitions is a film showing Poland and the world after 1945; the various fates of East and West, separated by an iron curtain; conflicts and wars; liberation movements in various parts of the world, including the events of December 1970,[11] and the Solidarity movement. I saw the emotions of many people watching the film; some of the visitors told me that this was one of the most passionate parts of the exhibitions. It usually takes a few hours of visiting, often with a feeling of fatigue and satiety, to reach the location of the film, and yet it still makes an impression. We tried to not unnecessarily saturate the exhibition space with sounds to avoid the cacophony occurring in many museums. However, at times, sound played a great role, as in the uprisings room, where in the immersive "underground" setting we hear whispers of praying people. In a room dedicated to the tragedy of prisoners of war, we played Orthodox Church funeral music, with a wall projection of a hecatomb of Soviet prisoners, over 3 million of whom lost their lives, mostly from starvation, while under German incarceration.

---

[11] Polish workers responded to a steep increase in food prices in December 1970 with a series of strikes, protests, and demonstrations, which were brutally suppressed by the police and army.

Some solutions we planned intentionally as attractions about which visitors would talk long after leaving the Museum. In the room devoted to breaking the Enigma cipher, we placed two multimedia stations next to the original German cipher machine. These contain reproductions of Enigma on which visitors can encrypt and decrypt a message.

The most important exhibition elements, however, were the authentic artifacts. Our entire team was passionately devoted to searching for them. When I left the Museum in April 2017, there were over 40,000 artifacts in the collection. More than 2,000 of them were displayed in the main exhibition. For comparison, there are a hundred or so at the Museum of the History of Polish Jews in Warsaw, where multimedia is the main medium of the narrative. The creators of this museum admit that it is a child of the 1990s, when they began to plan the exhibitions and when museologists, globally, gorged on the new technical possibilities.

It is impossible in this book to relate all the stories, or even all the most colorful ones, related to artifact acquisitions. Of exceptional significance, emotional as well as historical, was the memorabilia passed on to us by individuals. Usually connected to grandparents or parents, these objects, documents, or letters were often accompanied by detailed information about the circumstances of their creation and the fates of the people they represented. This was especially valuable because we were not only exhibiting artifacts but sharing the stories they told.

The breakthrough in this context was a nationwide campaign to collect memorabilia, which we carried out between October and December 2011. We promoted it widely in the media and managed to collect several thousand artifacts. The campaign had a snowball effect even after it ended. The Museum became known, and many donors contacted us on their own initiative. Of course, there were exceptions. Some donors had to be persuaded, even over several years; we had to gain their trust. Giving away the most valuable family relics is never an easy decision. One has to be convinced that they will be in good hands and used in a way consistent with the intentions of the donor, that they will not become part of a museum message that the donor would find unacceptable. Knowing some of the donors, as well as the circumstances under which the artifacts were donated, I was not surprised at all that so many of them reacted by announcing that they would withdraw the artifacts from the Museum when Minister of Culture Piotr Gliński announced the merger of the newly opened Museum of the Second World War with the Westerplatte Museum in 2016.

I will still mention some of the artifacts given by individual people, objects that have a special emotional or historical meaning for me. Others who worked on the exhibitions would probably mention other objects because we each experienced them personally and in a slightly different way. Joanna Muszkowska-

Penson donated the letters she exchanged with her parents when she was a prisoner in Ravensbrück. The same section of the exhibition also included an album made secretly in the same camp by Polish prisoners. Some of the drawings were the work of a well-known artist, Maja Berezowska, author of Hitler caricatures published in the 1930s, whom the Gestapo hunted down after entering Poland. The album was a gift to Andrzej, son of one of the prisoners, Marta Baranowska. He gave it to the Museum seventy years later. Both gifts became elements of the story of concentration camps.

Andrzej Stachecki gave us mementoes, including letters and secret messages from prison, from his father, Józef Stach, a forester shot by the Germans in the Piaśnica forests together with over 12,000 other prisoners. Because of these authentic artifacts, we could place Stach as one of the main characters in the story of the extermination of Polish intelligentsia in Pomerania. The family of Captain Antoni Kasztelan, a famous chief of Polish counterintelligence in Pomerania in the 1930s who was hated by the Germans, gave us the clothes and personal items that were returned to them after he was guillotined in the prison in Königsberg.

Olga Krzyżanowska donated memorabilia from her father, Aleksander Wilk-Krzyżanowski, commander of the Home Army in the Vilnius region. After the liberation of Vilnius, jointly by the Red Army and the Home Army, he and his soldiers were deceived and arrested by the NKVD and then sent deep into the Soviet Union. One item, a signet with an image of a wolf, reference to the commander's pseudonym, was made by his subordinates in the camp in Ryazan. Another personal item, a pair of glasses, had been with him in a prison cell in Warsaw's Mokotów district, where he died in 1951. In the exhibition, both artifacts are symbols of the postwar fate of Home Army soldiers.

We also received from the family of Witold Łokuciewski, the ace of Polish fighter pilots, a donation of his uniform and personal belongings. He fought in defense of Poland in 1939, then over France, and in the Battle of Britain in the famous No. 303 Squadron. In 1947 he returned to Poland, where he was persecuted by the Communist authorities. Because of these artifacts, we have a wonderful illustration of the fight of Polish airmen, and through the fate of Łokuciewski himself, we also talk about the repression that soldiers of the Polish Armed Forces in the West suffered when they returned to their country.

We also received a donation of the banner of the 6th Heavy Artillery Regiment of the Polish army, hidden in September 1939 after the capitulation of Lviv. After the war, at the time of resettlement of the Polish population to the west, the banner was secretly transported to Wrocław by a Polish family, who hid it there for almost seventy years. The family members declined to have their name disclosed.

A secret box hidden in the floor and used by the Home Army is another exceptional artifact. It had been kept in Warsaw's Powiśle neighborhood in the apartment of neurosurgeon Jerzy Choróbski. In the 1970s and 1980s he supported the democratic opposition and the Solidarity movement, which used the box in his apartment to store dissident publications. Professor Choróbski allowed us to remove the box; now it can be viewed in the Museum, and although it is in the section about the Polish Underground State, the visitor learns its entire history.

Once we know the provenance of artifacts, the stories related to them, and the people who donated them, an emotional map of an exhibition begins to emerge that is entirely different from what can be perceived at a first glance. The exhibition becomes something very alive and experienced personally, but also it evolves step-by-step as more artifacts are acquired. The artifacts came not only from strangers but also from members of our own team. One of our most valuable items, the previously mentioned handkerchief of Bolesław Wnuk, was a donation from the family of professor Rafał Wnuk, grandson of a rural activist shot by the Germans. Łukasz Suska, a great expert on the artifact market who had acquired many valuable objects for us (especially weapons), handed over the family souvenir himself. It was a bowl used by Jews whom his family, later recognized by Israel as Righteous Among the Nations, had hidden from the Germans. This gave us an astounding and very tangible sense of historical continuity, the creation of an exhibition by the descendants of those whose fates we presented.

Now and then an artifact came to us through a strange, circuitous route, a fact that conveyed something very important. The custodian of a Warsaw apartment building found in the garbage a pencil portrait of a girl with the inscriptions "To my daughter" and "Murnau." It had been drawn from memory by a Polish officer interned in a German prisoner of war camp. Apparently, the next generations did not find it valuable enough to keep it. Fortunately, Professor Zdzisław Najder, a literary historian, lived in the same building; the custodian gave this find to him, and he in turn gave it to us. In this case, a valuable artifact was saved and included in the section of the exhibition dedicated to prisoners of war. Yet many other objects were definitely lost. Moreover, our experience showed that the Museum was created at the last possible moment, when we still had a chance to obtain family souvenirs on a large scale. Many of them were accepted not so much for the main exhibition as to protect traces of the past; all are carefully preserved by the Museum and undergo conservation, and some may be shown in future temporary exhibitions.

The exhumation of mass graves was a very important source of artifacts. The part of the exhibition about the Volhynia massacre was built around objects excavated from burials. These artifacts had belonged to residents of two Polish villages, Ostrówki and Wola Ostrowiecka, who had been murdered by the Organization of Ukrainian Nationalists and the Ukrainian Insurgent Army. A historian, Leon Popek, son of one of the few survivors of the massacre, gave them to us. The objects are complemented by a film in which his mother, an eyewitness, talks about the course of events. This creates a sense of almost touching history, or at least getting very close to it. The Katyń display is also built around objects belonging to the victims and excavated from the burial. They are complemented with souvenirs and photos provided by the families of the murdered. We see the faces of the people, full of life, from before the war juxtaposed with what remains of them.

We also showed objects extracted during the 2001 exhumation of the grave of Jews burnt alive in a barn in Jedwabne in 1941. After acquiring them from the Białystok Institute of National Remembrance, which had conducted the now completed investigation, we did conservation work on all the items. We decided, however, only to display some of them in the exhibition, those which in our opinion were the most impactful. The keys found with the remains of the victims give this story an intensely human dimension—they had been hoping to return to their homes. Another recovered artifact was a statue of Lenin erected in the market in Jedwabne by the Soviet occupiers early in the war. In the first stage of the pogrom, the Jews were forced to topple this monument, which ritualistically stigmatized them as supporters of communism. Then, according to the accounts of a Jewish survivor, Szmul Wassersztajn, and the court testimonies of Polish witnesses and perpetrators, a group of several dozen Jewish men was forced to form a procession with the rabbi at the head and bring the toppled and broken Lenin to a barn on the outskirts of the town. The men were murdered there, right before the remaining Jewish residents were burnt in the same barn. Finding the same monument and including it in the exhibition was something extraordinary and special. There is no similar artifact in any other museum in the world, including the most important museums devoted to the Holocaust. It was also a very personal experience for me, as a historian who had dealt with the massacre in Jedwabne and coauthored the two volume *Jedwabne and Beyond*, a report on the subject of this crime and others perpetrated by Poles against Jews during the war.

We received many valuable artifacts from other museums, either as gifts or as loans. The Scandinavians were particularly generous. From the Military Museum of Finland in Helsinki we received equipment, including skis of a Finnish soldier, which we present in the part of the exhibition dedicated to the so-called Winter War, or the aggression of the Soviet Union against Finland in 1939 and 1940. From the Military Museum in Oslo, we obtained one of the most valuable gifts, also in financial terms, an original copy of the German Enigma encryption machine. The Oslo museum has one of the best collections of German military technology in Europe, since the Wehrmacht capitulated in Norway without a fight in the last days of the war.

Valuable artifacts were sourced not only from museums but also from other institutions not related to museology or history. The chief of police agreed that, if anywhere in Poland his subordinates confiscated illegally owned wartime weapons—which, it turned out, happened quite often—instead of destroying them by the standard procedures, they would pass them on to our Museum. Also, thanks to the Polish navy, we acquired a German torpedo fished, after seventy years, out of the mud of the Gulf of Gdańsk.

Perhaps our most spectacular achievement was the acquisition of a Sherman tank through an exchange of artifacts with two other museums. Bogusław Winid, the Polish ambassador to NATO, a historian and an acquaintance from my university years (later, as a representative of Poland to the United Nations in New York, he got us a typewriter used by the Polish government-in-exile in London), arranged for us to partner with the Royal Military Museum in Brussels. This museum was part of the Belgian army, whose artillerymen practiced firing on an American Sherman tank from the war. It was the Firefly model, on which production began in 1944. The 1st Armored Division of General Stanisław Maczek, who had fought at Falaise, liberated Belgium and the Netherlands, and advanced all the way to Bremerhaven in northern Germany, was equipped with this very model. The director of the Brussels museum proposed an exchange: we would get the tank wreck (there was no such model available in Poland at that time), and in exchange we would provide spare parts for a T-34 tank (including brake discs, a carburetor, and a gearbox) and a modern uniform and equipment (including a rifle) of a Polish soldier fighting in Afghanistan. We were somewhat surprised that the Belgians were so concerned with such artifacts, but we got them without difficulty and from completely legal sources. Perhaps Vladimir Putin was right after all when he said, trying to conceal the presence of Russian soldiers in the Crimea during the Russian aggression there, that any modern military equipment can be purchased in collectors' shops.

The museum in Brussels was so satisfied with the exchange that it threw in as a bonus a British howitzer from the Second World War. We did not need it, but the Museum of the Polish Army in Warsaw dreamed of having a howitzer. In exchange, they gave us the T-34 tank, which we needed badly.[12] As the main artifact at the end of the exhibition, it was to symbolize both the victory of the Red Army

---

12 It is relatively easy to get spare parts for T-34 tanks, as there were thousands of them in the Polish army; the tricky thing is to get a whole original tank.

over the Third Reich and the Sovietization of Central and Eastern Europe. In total, for one uniform and a rifle and several spare parts we obtained two tanks for the Museum. I joked that I would like to be able to make equally beneficial transactions in my private life.

We transported pieces of the Sherman to Poland on a trailer. Polish television TVN took some footage for a documentary about General Maczek's soldiers, who had wanted to liberate Poland, entering it on their tanks. They were not granted their wish in 1945, but now at least one of their tanks was "returning" to Polish soil. The tank was restored under the supervision of specialists from the Museum of Armored Weapons and painted in the colors of one of the regiments of the 1st Armored Division, becoming the main link in our Museum's story of the battles fought by Polish soldiers in the West. As part of the restoration, an old bus engine was put into it, thanks to which it could run, albeit only forward and backward, and participate in numerous historical reenactments, including D-day on Hel Peninsula. The tank ignited great interest among military fans who flocked to its every performance. I had the impression that they were almost euphoric about it and behaved like groupies at their idols' music concerts. When it was time, before closing in the ceiling of the underground gallery, for the Sherman to be installed in the exhibition (along with other oversized artifacts like the T-34, a railroad car, and the torpedo), its fans organized a protest action on Facebook, demanding that the tank be left on the surface. They even threatened to block the operation of lowering the tank, which was a big logistical challenge, into the underground museum building. The tank would continue its "political" life during the war against the Museum declared by Law and Justice, which will be discussed in the last chapter of this book.

We found many other artifacts for the exhibitions through our own efforts, often using information provided by various institutions and people. From the employees of the Polish consulate in Saint Petersburg, we learned about the graves of Polish deportees in the Arkhangelsk region in the north of Russia. The wooden crosses were rotten and would not last long. With the help of the consulate, we mounted new metal crosses on the graves, and the old ones we took to Gdańsk. After conservation they have become the most moving artifacts in the exhibition about the deportation of Poles to the East. Their counterpart, to some extent, was the door from the apartment of a Polish family expelled by Germans from Gdynia in 1939.

Another of the most emotionally touching artifacts was a prewar wheelchair used by patients of the psychiatric hospital in Kocborów, not far from Gdańsk; we learned that it was still in the hospital's basement. In our exhibition it became a symbol of the mentally ill and disabled who were murdered on a massive scale by the Nazis. In Kocborów in 1939 alone, Germans killed several thousand patients, both locals and those brought in from the Reich. When we learned that the Polish State Railways, as part of a renovation, intended to remove the wartime antiaircraft shelter that existed under train platforms in Gdańsk, we managed to obtain a door to it. It now is used in the section of the exhibition dealing with bombing.

Even at the last minute, when we obtained an artifact that was particularly valuable or suited to our story, we tried to put it on display. This required design changes. An exhibition is an extremely complicated creation, with hundreds of installations that are not visible to the visitor. Since it affects the surroundings, even a seemingly minor change creates a chain reaction. In 2015, we received Joseph Stalin's original pipe as a gift. In a gesture of appreciation the dictator had given it to a Kazakh officer, Bordżuan Momysz-Uła, who had shown particular bravery during fighting near Moscow in 1941. He had even been featured in a popular book in the Soviet Union, which described the defense of the capital. In the 1960s Momysz-Uła gave the pipe as a gift to the Polish writer and traveler Romuald Karaś, who had met him in Kazakhstan. Now the pipe came to us through Julian Skelnik, a Danish honorary consul in Gdańsk. Winston Churchill described in his memoirs how, during the Big Three conference in Tehran, Stalin used a similar pipe to indicate on a map where he wanted the borders of countries to be moved. Fortunately, we managed to find a place for an additional display case in the section of the exhibition talking about grand politics and forging the shape of the postwar world.[13] One of the last artifacts obtained that could still be displayed in the exhibition was a suitcase belonging to residents of Estonia deported to Siberia after the war, which was handed over to us personally in Gdańsk by the Estonian president.

We had to accept that we could use only a small portion of the collection of tens of thousands of artifacts for the opening of the Museum. Besides, we knew that volume is not the key to attract interest and elicit emotions in visitors. Sometimes one well-chosen artifact that illustrates a topic is more effective than a dozen that become somewhat repetitive. We amassed so many interesting objects depicting deportations of Poles deep into the Soviet Union and their everyday lives in exile that we had to resign ourselves to not showing some of them in the main exhibition. This created an opportunity to transfer artifacts to other museums. During the opening of the House of European History in Brussels, I was moved to see the shoes of Polish deportees to Siberia, made of straw and wood, which had been lent by the Museum of the Second World War.

---

13 For more about the pipe, see Marek Adamkowicz, "Fajka Stalina trafiła do Gdańska," *Dziennik Bałtycki*, March 7–8, 2015.

It has been pointed out that building a narrative exhibition resembles the creation of a film.[14] It has fast-paced action that engages the visitor, who will otherwise soon become bored. There are mood changes. Images play the key role: photographs, fragments of films from the era, statements by witnesses. A pioneer in this field was the United States Holocaust Memorial Museum in Washington, DC, opened in 1993, which created a model for many subsequent historical narrative museums. It is no accident, then, that after several years of work on that museum's exhibitions without satisfactory results, the task was handed over to the British film director Martin Smith.[15] Some commentators have pointed out that ultimately the exhibitions follow the structure of horror films, very popular in American culture. First, we see the birth of radical evil, with the Nazis taking power in 1933, and then its rise and the escalation of crime and horror. In the end, the evil is overcome by American soldiers, whom we see in 1945 as liberators of concentration camps.[16]

In our case, the creation of the exhibitions had much in common not only with film but also with a theatrical performance, in which scenography usually plays a significant role. The final effect resulted both from the importance we placed on authentic objects forming the bones of our narration and from the very sophisticated scenography prepared by Tempora. Almost every room had a separate design and atmosphere, incorporating different lighting and either sound or silence. Some rooms were deliberately claustrophobic; others (e. g., those presenting militaria) used the vastness of the entire building. The visitor encounters reconstruction of an air raid shelter in the London subway but also an artistic installation on hunger—empty metal plates—placed in the vicinity of photos from the Warsaw ghetto, Greece, and the Netherlands. In Greece about 100,000 people died of hunger; in the Netherlands about 20,000 perished during the last winter of the war. On entering the room dedicated to the atomic bomb, viewers are struck by the glow that prevails in this space, contrasted with the semidarkness from which they have emerged. Once their eyes adjust to the change in the light, visitors see, suspended above them, a full-scale replica of the bomb dropped on Hiroshima.

---

14 Mieke Bal, "Exhibition as Film," in *(Re)Visualizing National History*, ed. Robin Ostow (Toronto: University of Toronto Press, 2008), 15–43.

15 Edward T. Linenthal, *Preserving Memory: The Struggle to Create America's Holocaust Museum* (New York: Columbia University Press, 2001), 143–145.

16 Anna Ziębińska-Witek, *Historia w muzeach. Studium ekspozycji Holokaustu* (Lublin: UMCS, 2011), 189. See also Caroline Joan (Kay) S. Picart and David A. Frank, *Frames of Evil: The Holocaust as Horror in American Film* (Carbondale: Southern Illinois University Press, 2006).

Yet, behind all this, there are also books: hundreds of pages comprising successive versions of the exhibition scenarios, as well as thousands of books needed to prepare the first draft, verify all the facts, and document the history of the artifacts. So it is a monumental work, akin to arduous scientific research, although it is intended to bring about a different result, closer to a film or theatrical performance and appealing to a mass audience. All this must be coherent, interesting, and original, not overwhelmed by historical facts but at the same time verified, reflecting the latest state of historical research.

Then, the sheer scale of the exhibitions needs to be taken into account. Our main exhibition has an area of about 6,000 square meters, which makes it one of the largest in the category of historical narrative museums. By comparison, exhibitions at the US Holocaust Memorial Museum and the Museum of the History of Polish Jews measure approximately 4,000 square meters.

For many of us, regardless of what happened next, this was a fascinating, extremely enriching intellectual adventure that expanded our earlier understanding of history, which had been shaped mainly by academic experience. From the beginning, the work on the exhibition was coordinated by Piotr M. Majewski; eventually Rafał Wnuk began to play an increasingly important role. They were the only people who, even in the middle of the night or in another time zone (which happened during our business trips), were able to describe every corner of these huge exhibitions, using only the numeric identifiers assigned to individual modules, which they knew by heart. After a few years of work, each square meter was planned and devised in detail. Before the production and assembly of the exhibits in the museum building began, we walked through it virtually, using a specialized computer program that enabled us to move around in three-dimensional space, viewing cases and artifacts from various perspectives. We waited with growing impatience for the moment when, after all these years, we would see the final, finished exhibitions.

The Academic Advisory Committee, or a Multitude of Perspectives

The discussions held by the Academic Advisory Committee were extremely important in influencing the final shape of the exhibitions. It was a consultative body, which I had appointed personally, unlike the members of the Board of the Museum (later the Trust Council), who oversaw the operations of the Museum. The appointments to the latter were determined by the minister of culture. A galaxy of eminent historians from various countries sat on the Academic Advisory Committee. Its most senior member was Władysław Bartoszewski, who passed away in 2015, and who for us was a moral symbol: a former prisoner of Auschwitz, a member of the underground resistance group of the Council to Aid Jews, and after the war a member of the democratic opposition movement. As a historian, he was an expert on the Polish Underground State and the German terror. He did not live to see the Museum opened, but with great satisfaction—right before being fired—I succeeded in getting the square in which the Museum stands named for Bartoszewski.

Among the other Polish historians on the committee, Tomasz Szarota was an outstanding expert on everyday life in occupied Poland and other countries under German rule. Jerzy Borejsza had authored works on fascism and National Socialism translated into many languages, as well as books about Hitler's attitude toward Slavs. Włodzimierz Borodziej published a well-known book about German actions against the resistance movement in Poland and then many more works on both the Second World War and the postwar period. Jerzy Holzer, who died in 2015, had authored publications on interwar Poland and a best-selling history of Solidarity, first published by the underground, free press. More recently he published two very important books that directly related to the subject matter discussed in the Museum, *Europejska tragedia XX wieku: II wojna światowa* (European tragedy of the twentieth century: Second World War) and *Europa wojen 1914–1945* (Europe during the wars of 1914 to 1945). Published in 2005, the first book especially inspired me as I prepared the Museum's *Conceptual Brief*. It contained a very broad cross-sectional view of the war, covering many nations and especially the experiences of the civilian populations. Andrzej Chwalba from the Jagiellonian University was a true polyhistor; he discussed many subjects of the history of the nineteenth and twentieth centuries. Anna Wolff-Powęska, who joined the group later, dealt with historical memory, National Socialism, and Polish-German relations. Krzysztof Pomian, a philosopher we mentioned earlier, was a historian of ideas and museologist and undoubtedly one of Europe's most eminent intellectuals, publishing his books in many languages and countries. In the last dozen or so years, he has written extensively on the philosophy of history and museums, publishing, among others, a work on the beginnings of European museology in the modern era.[1]

A number of foreign historians specializing in the Second World War, Poland, and Central and Eastern Europe accepted the invitation to join the committee. Norman Davies was not only the most renowned researcher on Polish history in the world, but in recent years had intensely tackled topics related to the war. His book about the Warsaw Uprising was a best seller in Poland; in 2006 Davies published a monumental work, *Europe at War 1939–1945: No Simple Victory*. He drew attention to the significance of totalitarian ideologies, the fates of the civilian populations, and the specificity of war and occupation in Eastern Europe, not commonly known in the West. These were all topics that we planned to explore in the Museum we created. In addition to Holzer's book mentioned above, *Europe at War* was an important inspiration for me in thinking about the shape of the Museum's exhibitions.

Timothy Snyder published his famous *Bloodlands: Europe Between Hitler and Stalin* in 2010, when the work on the exhibition was already well under way. Nevertheless, his vision of the historical experiences of Central and Eastern Europe in the 1930s and 1940s was very close to how we planned to depict the terror of two totalitarian regimes even before the outbreak of war and then the crimes committed by the Nazi Germany and the Soviet Union once it had begun. In Snyder's account, civilians were also the main heroes and victims.

Israel Gutman, who died in 2013, was an outstanding scholar and a historical symbol, in a sense a Jewish counterpart of Władysław Bartoszewski. He fought as a member of the Jewish Combat Organization in the Warsaw Ghetto Uprising and then was imprisoned in the Majdanek and Auschwitz camps. He was the author of, among others, books about the Warsaw Ghetto Uprising and the Łódź ghetto. From the 1970s, he was associated with Yad Vashem in Jerusalem; for many years he managed its research center. He was editor of the seminal *Encyclopedia of the Holocaust* and *The Righteous Among the Nations*. Like Bartoszewski, Gutman had a personal attachment to our Museum, treating it not simply as a professional endeavor. I remember his appeals that the exhibitions be a warning and show what hatred and indifference lead to.

Ulrich Herbert, one of the most renowned German scholars of National Socialism, had authored a pioneering book about forced labor in the Third Reich and studies on Nazi terror and the power elite, including a great biography of

---

[1] Krzysztof Pomian, *Collectors and Curiosities: Paris and Venice, 1500–1800* (Cambridge, UK: Polity Press, 1990).

Werner Best, a founder of the Reich Main Security Office. Henry Rousso was one of the most outstanding French scholars of the Second World War; he was the author of seminal works on Vichy France and the postwar assessment of it. Pavel Polian, a Russian historian and demographer, published, among others, a work on deportations in the Soviet Union, as well as books on forced laborers and prisoners of war and on the Holocaust in the occupied territories of the Soviet Union.

Since, from the start, forced labor, deportations, and the fates of prisoners were supposed to be an important part of our exhibitions, Herbert's and Polian's expertise was very important. A great advantage of the latter was the fact that he was not associated with any Russian government institutions; he was a professor at the Russian Academy of Sciences, he had spent a lot of time in the West, and he had published some of his books there. Polian introduced the Russian perspective to our discussions, but without the political entanglement and instrumentalization of history, which ever intensified under Vladimir Putin's rule. Elie Barnavi, already mentioned, participated from the beginning in the committee's meetings, for the first years appearing on behalf of Tempora, which was preparing the exhibitions, and later as a full member of the group.

It was an honor for me and my colleagues that these eminent historians and intellectuals from Poland and abroad accepted the invitation to participate in the creation of the Museum. They brought their knowledge, experience, and professional contacts, as well as different national perspectives, which laid the ground for fascinating discussions and gave us the chance to develop world-class exhibitions. It was obvious to me that since the Second World War was a global, not only Polish, experience, scholars from various countries should join the Academic Advisory Committee. In any event, this was a standard approach in other important museums and cultural institutions. Still, at the beginning, this approach caused some people serious doubts and concerns. A right-wing journalist expressed concern that foreign historians would impose solutions in accordance with the interests of their nations. "And if a German historian on the committee accuses you of hurting his nation?" he asked.[2] This represented not only anxiety about the "Polishness" of the Museum but also a purely political perception of the exhibitions and the work on them.

Many times I was later accused of not having, or insufficiently including, on the committee historians associated with the political Right. Piotr Semka regretted that only two of the committee's members, Tomasz Szarota and Andrzej

---

[2] "Pokażmy światu wojnę w jej pełnym wymiarze," conversation between Piotr Zaremba and Paweł Machcewicz, *Dziennik*, November 27, 2008.

Chwalba, could be regarded as representatives of a more "traditionalist" and "patriotic" approach to history, while the others had a liberal orientation.[3] I do not know if Szarota and Chwalba enjoyed the qualification given to them by the journalist; I, in any case, avoided using a similar label. I was not interested in the political views of the invited. It was important to me that they were outstanding historians, expert in the war, twentieth-century totalitarianisms, and the Holocaust, and that they possessed knowledge of historical memory and museums. It was important that at least half of them (Barnavi, Borodziej, Chwalba, Davies, Herbert, Pomian, and Rousso) had previously sat on academic councils of many historical museums; some were involved in creating exhibitions. They knew that the exhibitions could not simply reflect the information contained in books or encyclopedias.

The Academic Advisory Committee met twice a year. Initially, it discussed the overall concept of the Museum, then the thematic and spatial arrangement of the exhibitions, the arrangement of the rooms, the amount of space devoted to specific topics, and the individual parts of the exhibition. Finally, we presented to the members of the committee specific, detailed solutions, such as the most important films or multimedia presentations. I remember long discussions about the place in the exhibitions of forced resettlements, their definitions, and the terminology used in relation to them. According to the position of the committee, proposed by Włodzimierz Borodziej, we used three notions, depending on the course, legality, and scale of coercion: resettlement, deportation, and expulsion. During the discussion on the section devoted to the Holocaust, the main interpretive axis finally emerged, giving three perspectives—perpetrators, victims, and witnesses—following the classic view of Raul Hilberg.[4] Norman Davies insisted that in the English text we use not Hilberg's word "bystander" but the broader word "witness," which also connoted less passiveness. A witness could be a person who behaved in a completely passive way, or a person who helped Jews, or one who took part in their persecution or benefited from their tragedy—for example, by taking over their property. This showed how important it was to have foreign historians on the committee, whose linguistic sensibility was irreplaceable. We also decided not to use the concept of "total war," which was tainted by a strong association with Nazi vocabulary. Joseph Goebbels had called the Germans to "total war" in his famous speech delivered in Berlin in February 1943. Instead, we adopted the term "war of annihilation," which em-

---

3 Piotr Semka, "O co chodzi w sporze o Muzeum?," *Do Rzeczy*, July 25, 2016.
4 Raul Hilberg, *Perpetrators, Victims, Bystanders: The Jewish Catastrophe, 1933–1945* (New York: HarperCollins, 1992).

phasized the unprecedented brutality of this conflict, especially against civilian populations. I also remember us all watching the film being prepared about the bombing of cities, which began with German attacks on Polish cities but in the final part showed Allied air raids on Germany and the American raids on Japan. With such a difficult topic, every detail counted, even the background music of the film. On viewing the first version, the committee members perceived the change of music at the moment when the Allies began to bombard Germany as too joyful and triumphant. The idea was to show clearly who began this method of waging war, but we certainly did not want to suggest enthusiasm for the deaths of hundreds of thousands of German and Japanese civilians under Allied bombs.

The clash of different life experiences, as well as generational, professional, and national perspectives, was very interesting. The German historian thought our exhibition was too emotional and our choice of media baroque, but at the same time, after seeing the multimedia presentation on the Stalinist terror, he stated that he had not seen such an evocative and interesting presentation in any Western European museum. Such assessments confirmed our conviction that we had adopted the correct method of translating the Eastern European experience for the West.

Contrary to the predictions of right-wing columnists and politicians, the nationality of the members of the committee did not mean that they wanted to influence the shape of the exhibition in accordance with their raison d'état and national interest—for example, to minimize their own nation's responsibility for crimes or for causing the war. Here is a relevant fragment I pulled from the committee minutes for January 30, 2012: "In his remarks, Prof. Ulrich Herbert drew attention to how necessary it is to emphasize the fact that a large part of the Germans responsible for the extermination of Jews belonged to a very well-educated social group." The idea that a historian must first present the raison d'état and pursue the historical policy of his government, and not seek the truth, has become a dogma of the political Right in Poland. Fortunately, politicians' expectations are fulfilled only by the minority of our community, and to scholars from most countries they are pure abstraction, negating the basics of professional ethics. From the very beginning, the committee debates embedded the conviction that our Museum of the Second World War project had a unique character among other existing museums. Here is a fragment of the discussion at the first meeting in June 2009:

> Professor Ulrich Herbert stressed that the idea of creating a Museum of the Second World War within the *Conceptual Brief* framework presented to the committee is difficult to implement, innovative, and unprecedented, because the topic of the Second World War has al-

ways been presented through the prism of national history. . . . After the break, referring to previous statements of the members of the committee, Paweł Machcewicz explained that, yes, creating a European narrative about the Second World War is an unfeasible task. One can, however, attempt to break away from strictly national perspectives through a comparative approach within individual thematic blocks.

At the beginning of our work in 2008 or 2009, I could not, of course, have predicted the change in the political atmosphere in Poland, many other European countries, and the United States (Donald Trump's victory) that occurred in the years that followed—an increase in anti-European, nationalistic, and even xenophobic sentiments; fear of "the other"; isolationism and closure within one's own national community—all of which also influenced perceptions and approaches to history. Yet, when I now think about our beginnings, knowing everything that happened later on, I am strongly convinced that we made good use of a unique and possibly the only moment to create a museum that not just talks about the Polish experience but also presents the fates of other nations and the war as a transnational, universal experience. In Poland in 2007 and 2008, there was still optimism about our presence in Europe, the benefits of that, and the need and opportunity to tell the world about our experiences, which should become a part of a wider and reshaped European historical memory. It was probably a measure of our confidence at the time—I'm not just talking about myself and my colleagues but in a much wider sense—the conviction that we could create in Poland a unique museum on a European and global scale.

I also think that the Museum of the Second World War was a generational endeavor, created by people who had already been professionally shaped in a free, democratic Poland, with pro-European aspirations. At the same time, at least those of us fairly familiar with Europe and the world who knew how low the level of knowledge about our history in the West was, believed that it was possible to change this state of affairs. The museum, although it opened in 2017 in a completely different era, was a child of that time, which ended after the outbreak of the financial crisis, the intensification of anti-European populism, and the conviction that the European Union, and with it Europe, had entered a deep crisis. I believe that if I had proposed creating the Museum of the Second World War a few years later, the idea would not have been implemented. Politicians, not just on the right, would not have looked at it favorably; I would not have received so many public resources. In all probability I would not have introduced such an idea at all under these changed circumstances.

The change in atmosphere between when we started our work in 2008 and when we ended it a few years later is perhaps best captured in the film presented in the last room of the exhibitions, devoted to the long-term effects of the war. In our original plan, this montage of footage and photos showing the most important world conflicts after the Second World War, as well as the different lives of millions of people on both sides of the Iron Curtain, was to take us to the great turning point of 1989. Up to that year, the images appeared on two adjacent parts of the wall: the East on one side, the West on the other. Images from 1989 and after were combined and showed scenes from the collapse of Communist regimes in Central and Eastern Europe, the formation of the first democratic government led by Tadeusz Mazowiecki, the destruction of the Berlin Wall, the collapse of the Soviet Union, and the withdrawal of Soviet troops from our part of Europe. We wondered whether, at the end, we should allude to the entry of Poland and other post-Communist countries into the European Union. Essentially, the message closing the exhibition was to be optimistic, illustrating the triumph of freedom and democracy, overcoming the division of the world and the enslavement of Poland and Central and Eastern Europe, the longest-term consequences of the Second World War. However, during the eight years of work on the exhibition, the world around us had changed, and not much was left of the earlier optimism and belief that everything was going more or less in the right direction. That was why we decided to bring the final film to the present day. The last scenes would show the viewer the wars in Syria and Ukraine, the ruins and suffering civilians in Aleppo and Donbas, and the refugees.

The goal was to make clear at the very end of the visit that the world did not close the book on war, violence, and suffering in 1945. They surround us today as well, and the inclination to violence is in us. In this way, the exhibition's finish lost its original optimistic overtone but became, and with it the whole Museum, much more universal and true. It is also symbolic that the first change introduced to the permanent exhibition, after the Law and Justice Party took control of the Museum in April 2017, was removal of this film. However, several hundred thousand visitors managed to see it. It will also probably be shown many times in various places as an illustration of modern censorship. I also hope that someday it will be displayed again in the Museum itself. Anyway, only a few hours after it was removed, the film was available on the Internet, placed there by an anonymous visitor who recorded it with his or her smartphone, proof that censorship in contemporary times has little effect.

From the Concept to the Exhibitions and Everything in Between

In the creation of large and multithematic exhibitions, one has to reconcile various, often seemingly contradictory goals. On the one hand, one should remember the initial assumptions that defined the goals, character, and essential content. On the other hand, new ideas and threads appear when detailed scenarios and design are being prepared, during the dialogue between historians and designers, during the acquisition of photos and artifacts, and through the recording of recollections with witnesses of history. Some of the earlier ideas and themes seem unnecessary or impossible to implement. An illustration of the latter case may be the change of Tempora's original design of the section on the resistance of occupied European nations. Most of the galleries in this exhibition convey the impression of underground rooms and cellars. This is to underline the conspiratorial nature of most of the resistance movements, especially the Polish Underground State, which dominates the story in this section. Above these rooms, there was to be one more level, possible due to the overall fourteen-meter height of the main gallery, which visitors would not enter but would see from below. It would contain a reconstruction of a street in the occupied city, with figures of people, lanterns, and even a car. This was to reinforce the impression that seemingly normal life was happening on the surface, but in reality hundreds of thousands of people took risks and were involved in efforts to regain their freedom and liberate their countries. It turned out that mounting the second, upper level of exhibition was technically very difficult and troublesome from the point of view of fire-safety requirements, and it was not clear whether this higher level would be easily seen from below. We replaced the planned reconstruction of the street with screens on which scenes of everyday life, from various occupied cities, constantly shifted in a montage of film shots from that time. In my opinion, the result is better, much more authentic than what we could have achieved with even the best-executed conventional design.

In some cases, we supplemented the exhibitions with thematic threads or even entire sections that we had not anticipated before. About halfway through the work we became aware that the exhibitions contained so many disturbing and difficult messages—in the sense of information and images—and were so overwhelming in their sheer size that they could be too difficult for children to absorb. We decided that a separate space should be created for visitors under the age of twelve. This was Rafał Wnuk's idea. We carved out about four hundred square meters from the huge temporary exhibition hall for the exhibition *Journey in Time*. It tells the story of the war and occupation through the prism of the fate of one middle-class Warsaw family. The family had relatives in the territories incorporated into the Reich and in the East, which allowed us to create a wide panorama of Polish fates. The backdrop of this story is a changing apartment in a Warsaw tenement house at three different moments of the war; through the window we watch street scenes and see how the city has changed—here the multimedia presentation was invaluable.

Corrections in our plans concerned not only technical and design solutions but sometimes also deeper assumptions. We assumed that the main exhibition would have a primarily thematic rather than chronological layout, which would give greater flexibility in showing selected aspects of the war experience.[5] For example, the *Occupations, Terror, Resistance* thematic layout remained the skeleton of the Museum, but as we worked, we quickly concluded that it must be placed in a clearly defined chronological framework. Without it, the viewer would feel lost, and many fundamental issues, such as the evolution of German occupation policy and terror, could not be explained. Ultimately, the visitor is consistently guided by the Museum's narrative from the First World War and the revival of the Polish state in 1918, through the birth and expansion of totalitarianisms in the 1920s and 1930s, through the war in its subsequent scenes and military campaigns, up to 1945, when the visitor sees the end of occupation and military operations, decisions of the victorious powers on a new shape of the world, and the prosecution of war crimes; the last hall leads her or him back to the present day.

Initially, we assumed that the Polish stories would total half of the exhibitions, but we quickly learned that this was completely unrealistic as many Polish wartime experiences could not be separated from the history of other nations. Is the Ribbentrop-Molotov pact primarily a part of Polish history because it decided Poland's fate, or is it broader, because it marked a turning point in the outbreak of war and the division of many Eastern European countries between two aggressors? In the same way, it was impossible to distinguish the extermination of Polish Jews from the entire Holocaust—though they did account for about half of its victims. Should the stories about concentration and forced labor camps be arranged according to purely national criteria? There were more stories in the Museum about Poland than about other countries participating in the war; some parts of the Museum were entirely devoted to it, like the Warsaw street of the

---

5 Already in 2008 the right wing was attacking us for this approach, as it allegedly hid the fact that Poland was the only country fighting from the first to the last day of the war. It was axiomatic that they searched for hidden, sinister intentions in every word we said, and everything, even the most innocent, could become the subject of the most serious suspicions and accusations.

1930s or a large section occupying several hundred square meters dedicated to the war of 1939. Other parts are dominated by Polish themes: the Polish Underground State, probably the most detailed part of the Museum, occupies the most space in the resistance section. In the gallery on anti-German uprisings in occupied countries, the most attention was devoted to the Warsaw Ghetto Uprising and the Warsaw Uprising of 1944. In the military section, visitors learned mostly about the fight of Polish soldiers and saw weapons and uniforms mostly belonging to them, which proves that the stubbornly repeated accusations that we had forgotten about the heroism of Polish soldiers were a total lie. The essence of our narrative was the comparison of the fates of Poland and other countries, which led to an understanding of both their differences and their similarities. This was, for example, how the story of the German occupation of conquered countries was constructed. Such a comparative approach proved its strength in the depiction of anti-Jewish pogroms in 1941. Jedwabne was shown next to Lviv in Eastern Galicia, Kaunas in Lithuania, and Jassy in Romania, which allowed a much better understanding of the mechanism and causes of pogroms.

Our starting point was to emphasize the fate of civilian populations, and that actually constituted the dominant part of the exhibitions. During our work, however, we obtained a huge number of interesting and sometimes unique military artifacts: weapons, uniforms, and items belonging to soldiers. It would be a sin not to show them, especially since they often stirred the greatest interest, especially in young men. Perhaps the Museum's most famous artifact was the already mentioned Sherman tank in the colors of the 1st Armored Division of General Stanisław Maczek. We created two large rooms entirely devoted to the military dimension of the war, in which all campaigns were presented in chronological order. However, we made sure that this was not just a chronicle of battles. We showed all aspects of a soldier's life, beyond the fighting: from mobilization and training, through free time, longing for loved ones, and captivity, to wounds and death. We also showed a broader socioeconomic dimension of the war: the mobilization of economic potential, the work of women in the armaments industry, and technical inventions.

One of the biggest intellectual challenges was to create a coherent story about the terror of the occupiers. This block was the largest and probably the most important part of the Museum. It began with a detailed presentation of German terror against Polish elites. Its greatest intensity took place in Pomerania, but in this section we also showed mass executions and the massacres of entire towns, including the Wola district in Warsaw in August 1944, but also Lidice, Oradour-sur-Glane, and places in Greece, Italy, and Belarus (in this last country, terror was particularly extensive). The visitor went through a room with the already mentioned Bolesław Wnuk's handkerchief, showing the parallel between

German and Soviet terror directed against Poles, to the gallery dedicated to the Katyń massacre. From there, the visitor's path led to forced resettlements, both German and Soviet. This was an important section because it showed that the Nazi Germany and the Soviet Union began the great social engineering and resettlement of millions of people, not the Allies or the Poles in 1945.

Next, there were forced labor and concentration camps. The design of that section alluded to the barracks and showed, above all, the various aspects of the prisoners' fates: from the violence they constantly encountered, through lethal work, diseases, medical experiments, and the conspiracy networks created by inmates, to death, symbolized in the camp reality by a crematorium. Talking about camp conspiracies, we depicted Captain Witold Pilecki, who created the first such network in Auschwitz, and, relatively unknown to this day, Helena Płotnicka. A mother of six children living near Auschwitz, she smuggled food, medicines, and letters to the prisoners, for which she herself was imprisoned in the camp, where she later lost her life. For me, this was an example of heroism of the highest order and, at the same time, heroism unnamed. We wanted to restore the memory of exactly such people in the Museum.

Yet another display showed the murders committed by the Nazis on several hundred thousand disabled and mentally ill people—a category of victims to a certain extent forgotten and recognized only in the last dozen years or so. We reminded visitors that these murders began in the autumn of 1939, largely on the Polish lands incorporated into the Reich, including Pomerania. This story introduced the next section: the Holocaust. The link between the two parts was not only the intention to eliminate whole groups of people deemed unnecessary or harmful but also the fact that the centers of extermination and the gas chambers for the mentally ill and disabled were manned by the same people who were then sent to the General Government, where in 1942 they created extermination camps for Jews.

The Holocaust was first recounted in a prologue showing various experiences and perspectives—victims, perpetrators, and witnesses—and then was told in three chronologically constructed parts. The first depicted the events of 1941, starting with the extermination of Jews in the East by the mobile deployment groups (*Einsatzgruppen*) of the German security police. Here were also shown the previously mentioned pogroms carried out by the local populations, in many cases, as in Jedwabne, instigated by the Germans. The second part dealt primarily with the extermination of Polish Jews. The fate of Jews in the Łódź ghetto was also presented here. We chose it consciously, because the story of the Warsaw ghetto was shown in detail at the Museum of the History of Polish Jews. The next section related the extermination of Jews throughout occupied Europe, which took place primarily in Auschwitz. Here, too, was depicted the murder of the Roma and Sinti. The closing gallery was a monument to the victims—visitors saw on glass panels photographs of the faces of several hundred victims from numerous countries. Close by, there was a section dedicated to ethnic cleansing. We showed the murder of about 100,000 Poles by Ukrainian nationalists in Volhynia and Eastern Galicia and the extermination of several hundred thousand Serbs, Jews, and Roma at the hands of Ustašas, the Croatian fascists. These were not crimes committed by either the Germans or the Soviets, but they would not have happened without the destructive processes the Nazis and the Soviets launched, without the example they provided of eliminating entire ethnic, racial, and social groups.

The innovative approach in this part of the exhibition lay in incorporating the Holocaust into a broader picture that took into account the whole of the German terror machinery. In many museums it is shown separately, often as the only or at least the dominant theme, such as in the United States Holocaust Memorial Museum in Washington, DC, Yad Vashem in Jerusalem, the Mémorial de Shoah in Paris, and many other less known Holocaust museums. The German terror affected many groups of victims, who were imprisoned, forced into slave labor, and deported. Among them were Jews, who before being murdered were often forced laborers, prisoners of concentration camps, and forcibly resettled in ghettos. Such a comprehensive approach created a unique opportunity to show the interdependence between various elements and directions of Nazi violence and genocide, the evolution of German occupation policy toward various social, national, and racial groups, and finally the differences and the shared fate of the victims.

The symbol of the latter was the railway wagon placed in the middle of the section dealing with the terror. From it, the visitor went to all the sections described above. This unique artifact had an extraordinary story that is worth including here. Produced in Germany during the First World War, it was taken over by the reborn Polish state and in September 1939 by the Soviets after their aggression against Poland. In 1941, after the Third Reich's attack on the Soviet Union, it fell into the hands of the Germans, who used it till the end of the war, when it was recovered by the Polish State Railways. In such wagons, the Soviets transported hundreds of thousands of Polish citizens to the East, and the Germans transported deportees, forced laborers, and prisoners to concentration camps and Jews to extermination camps. The wagon also illustrates the power of the artifact in the exhibition; its story tells us not only about the occupation terror and the Holocaust but also about the emerging and failing states in Central and Eastern Europe in the twentieth century and their changing borders. It should be noted that in the US Holocaust Memorial Museum and Yad Vashem, similar wagons symbolize the fate of Jews only. We deliberately decided to present the wagon in a more universal way, which by no means intends to portray the fates of all victims as equivalent—they were not at all.

The fact that we devoted so much space to the terror of the occupiers and crimes against the civilian population and prisoners of war led to accusations even before the Museum opened that we presented a one-sided view of war with definitely pacifist and antiwar leanings. This view was too "dark," focusing on the monstrosities of war. There was allegedly too little room in it for heroism and combat, especially by Polish soldiers. I will cover this in detail further. However, it should be emphasized that from the very outset, we wanted to show first of all the everyday experiences of ordinary people, both soldiers and civilians. The civilian perspective was the most important, because civilians were not only the main victims of this war but also capable of the greatest sacrifice and heroism. The departure from the militaristic perspective was also due to the conviction that today's visitors, in the great majority, would not have experienced military service, which was not mandatory in Poland and in the vast majority of European countries. It would be much easier for the visitor to identify with ordinary people and not with people in uniform. A lot of attention was, nevertheless, devoted to soldiers in this huge and multilayered Museum, although not necessarily to generals and marshals but rather to their subordinates.

We also tried to encourage independent reflection. Exhibitions showed controversial issues, requiring visitors to come up with their own answers about human attitudes, choices made, and the price of resisting, which often resulted in massive and violent repressions, impacting (often especially) those who had no defenses and just wanted to survive. The manner in which we depicted the Warsaw Uprising was an example of this approach. The visitors experienced it from a cellar, like the cellars in which thousands of civilians took cover during the uprising from air raids, artillery fire, and German troops. Visitors saw joy on the streets of the liberated part of Warsaw, boys and girls from the Home Army preparing to fight, and the liberation of the telephone exchange building (one of the insurgents' greatest triumphs), as well as the killed, the wounded, and the ruins. They heard the prayers of terrified people asking God for salvation and learned about the growing grief among civilians as the monstrous losses mounted, the hopelessness of the battle, and even the aggression against the insurgents, blamed for what had happened. This is a true picture, attested to in hundreds of memories and accounts and recorded in the famous *Diary of the Warsaw Uprising* of Miron Białoszewski[6] but virtually absent in the Warsaw Ris-

---

6 Miron Białoszewski, *Pamiętnik z powstania warszawskiego* (Warszawa: PiW, 1970).

ing Museum, which focuses almost exclusively on the heroic and, above all, military dimension of the insurgent uprising.

Does such a broadening of the perspective, fully consistent with the historical record, harm the image of the uprising? Or does it weaken the nation, the strengthening of which, according to many right-wing politicians and journalists, is the main task of historical museums? In my opinion, the answer is no. The national and civic community should be built on the authentic experience of the past in all its complexity, because otherwise it becomes only a facade, a feeling experienced only during special celebrations. For me, the role of a museum is not to convey the one and only, uncontested truth but rather to encourage visitors to reflect on themselves, on the decisions and choices of people whose fates they see in exhibitions, and also on how they would have behaved in similar circumstances. For this reason the Museum covers extensively the story of the Polish Underground State, which was constructed to a large extent from the grass roots up, by civilians, as a result of spontaneous activism and reflected all the diversity and internal conflict of Polish society.

We showed Irena Sendler, who coordinated the rescue of Jewish children from the Warsaw ghetto, organizations such as the Council to Aid Jews, and a pharmacy, Apteka Pod Orłem, that operated in the Kraków ghetto. We also devoted attention to those individual people who were saving Jews without any institutional support, which was certainly the most difficult task. However, we also presented the crime in Jedwabne and talked about other places where Poles killed Jews, as well as about indifference or hostility toward them.

The most important issue was to keep the correct proportions in showing these various attitudes, consistent with the historical truth, which, of course, is always a subject of discussion and dispute. Even at the beginning of the 2000s, during the debate on Jedwabne alone, two visions of how to present the national past were outlined. Andrzej Nowak believed that the nation should be constituted from references to heroism and martyrdom and not to the dark pages of its own history. "As a community we can confront the monument to the heroes at Westerplatte and feel pride. At the monument in Jedwabne, in our shame for what happened there, we can feel no unifying pride," wrote the conservative historian from Kraków.[7] I had responded to him, when I was still director of the Public Education Office of the Institute of National Remembrance, that it was our duty to face all the pages of our history, constantly:

> Discovering the whole truth about our past, restoring the full splendor of things of which we should rightly be proud, as well as tearing the veil off shameful matters that occur in the

---

[7] Andrzej Nowak, "Westerplatte czy Jedwabne," *Rzeczpospolita*, August 1, 2001.

history of each country, is the best way to build a nation rooted in the past, aware of the generations before it. The community that is not afraid to confront its own history will not be surprised by discoveries that endanger its perpetuated self-image.[8]

Bringing all this back to the museum context, I did not think that visitors should leave our exhibitions refreshed and uplifted after watching the story of the heroism of the war. Rather, I wanted them to come out of the Museum moved and even shocked by what they had seen, trying to sort it all out in their thoughts, to give it meaning in their own consciences, to perhaps return to the exhibition again to seek answers to the questions troubling them. I was to see over time that this was indeed the reaction of so many visitors with whom I had a chance to speak.

My view of the role of the Museum did not go as far as that of art historian Piotr Piotrowski, former director of the National Museum in Warsaw, who in his manifesto, *The Critical Museum*, proclaimed that museums should be iconoclastic, should provoke, should display primarily controversial topics that disturb accepted ideas.[9] I think that the role of the historical narrative museum is different: first of all, in my view, it must present everyone with a reliable, true story about the past, while, through its selection of themes, individuals, and artifacts, providing its own interpretation.

In this regard I would rather refer to the classic reflection on the role of museums introduced by Canadian museologist Duncan F. Cameron in the early 1970s. In his famous article "The Museum: A Temple or the Forum," Cameron rejected the model of the museum formed in the nineteenth century, which transmitted an agreed-upon, codified truth about art, history, or other areas that the visitor was to accept and internalize without asking unnecessary questions. It was a one-sided model of communication with visitors, whose role was essentially passive. The museum was a "temple" transmitting the revealed truth, which was not discussed and not subject to debate. According to Cameron, if museums wanted to be living places that attracted audiences, they must become "forums," spaces for debate, reflection, and a clash of conflicting views.[10]

This vision of the museum was closer to mine, although, of course, the specificity of the historical narrative museum meant that the visitor encountered an already prepared story. Still, it was important that the story left room for various

---

[8] Paweł Machcewicz, "I Westerplatte, i Jedwabne," *Rzeczpospolita*, August 9, 2001.
[9] Piotr Piotrowski, *Muzeum krytyczne* (Poznań: Dom Wydawniczy REBIS, 2011).
[10] Duncan F. Cameron, "The Museum, a Temple or the Forum," in *Reinventing the Museum: Historical and Contemporary Perspectives on the Paradigm Shift*, ed. Gail Anderson (Walnut Creek, CA: AltaMira Press, 2004), 61–73.

interpretations, showed the whole variety of attitudes, and related choices and dilemmas that were not always clearly black and white. This was also true of the language in the exhibition text; we tried not to overload it with strong value judgements. Visitors have a right to draw conclusions from what they see. Our opponents on the political Right, however, had a completely different vision of the Museum as transmitting unambiguous truths, essentially serving as a propaganda tool for modeling minds according to a top-down vision in which there was no room for questions and doubts or even one's own interpretation—essentially as the "temple" and not the "forum," where the debate goes on.

Before the Storm

By 2015 all our planning had come together to form a coherent entity. The detailed design had been refined and could only have been subjected to corrections and changes before final production and installation in exceptional cases, as when we acquired a new artifact of outstanding historical value that fit our story. Building construction was in its final phase. Walking around the interior of the building, we could already imagine exactly what we would see in individual galleries; some of the oversized artifacts were already in place.

As the date of the expected opening approached, promotion of the Museum was becoming more and more important. This was all the more necessary because our activities were constantly contested and distorted by people close to the Law and Justice Party, who repeated the mantra that the Museum was "not Polish enough" without even bothering to become familiar with its actual shape. We made a movie visualizing the exhibitions and also showing some of the most interesting artifacts. Maja Ostaszewska, one of the most famous Polish actresses, was the narrator. I chose her because of her roles in the movie *Katyń* by Andrzej Wajda and the *Time of Honor*, a very popular TV series about the Polish wartime experience. However, she was a very versatile actress who also performed in productions for more general audiences. Maja Ostaszewska also became the voice of the Museum; she recorded the audioguides. Recounting the war in the voice of a woman demolished stereotypes and reinforced that ours was not a strictly military museum but showed the fate of civilians, women, and children.

I tried to interest the government in the Museum's fast-approaching opening. As a member of the Historical Council of the Ministry of Foreign Affairs, at one of the meetings I proposed to invite the leaders of European and world countries to the opening, which would provide a great opportunity—even greater than the celebration at Westerplatte in 2009—to showcase Polish heritage abroad. After all, this was one reason why the Museum had been created and so much of Polish taxpayers' money spent on it. Minister Grzegorz Schetyna expressed interest, but it was not a done deal. The forthcoming presidential and then parliamentary elections had eclipsed all other topics. However, I felt that throughout the years, we were able to build such an intense network of contacts with the most important world war museums that the opening of the Museum of the Second World War could be a big international event, even without the support of the government. In any event, we could already see on the horizon the end of the work that had started in 2008. But I would never have predicted at that time that everything would soon be turned upside down and that the Museum and I would become public enemies of the new political order.

III The War

"The Polish Point of View"

Signs of the approaching storm could clearly be seen and heard long before it broke, beginning with Jarosław Kaczyński's attacks on the Museum in 2008, just after the release of our *Conceptual Brief*. The leader of Law and Justice did not forget about us in the following years. The statements that he expressed about the Museum at the party's congress at the end of June 2013 echoed widely. He announced that his party wanted to "change the shape of the Museum of the Second World War so that the exhibitions in this museum would express the Polish point of view."

In response, I issued a statement to the Polish Press Agency that was published by all major media. I pointed out that the leader of Law and Justice "suggests that, currently, the Museum I am managing does not represent the Polish point of view, but the point of view of other nations. Which nations? The President of Law and Justice has not clarified." I wrote further,

> I would like to emphasize, although it seems obvious, that the Museum of the Second World War is a Polish cultural institution, formed by the minister of culture and prime minister of the Polish government, financed from the Polish budget. The Museum ... is created by Polish historians and museum professionals. Its main goal is to introduce the Polish and Central European perspective to the world's narrative and historical memory, within which our experiences often take up little space and are not sufficiently known and understood. The Museum intends to change this state of affairs. And the best way to do this is to show the Polish experience in a historical context, against the background of the experiences of other nations and of the entire European continent and the world, because the Second World War was a global conflict. Such a view does not pose a threat to "Polishness"; on the contrary, it allows a better understanding of the specificity of Polish history. We do not have to be afraid of historical dialogue with other nations and of cooperation and exchange of views with historians and museums in other countries.

I also encouraged a debate that would address the Museum's actual actions and plans, not perceptions and allegations detached from reality, which as a rule was what our opponents had presented from the very beginning.

> Although the Museum is just being created, we have already published many books, mounted a number of exhibitions, including a permanent outdoor exhibit at Westerplatte, organized professional conferences, and offered extensive educational activities. It is our concrete output that should be the basis for the assessment, including criticism, of the Museum's current activities. It is important to exchange factual arguments. The suggestions that the Museum of the Second World War does not represent the Polish point of view certainly do not belong among these arguments.

I concluded,

> It is also worth emphasizing that nobody has a monopoly on judging what is the Polish point of view; Poles differ among themselves, and this is one of the greatest values of an independent, democratic state, which we have enjoyed since 1989. Worrisome also is the announcement that a political party, after taking power, will change the content of the Museum's exhibitions. This strikes directly at the principle of autonomy of cultural institutions. Politicians interfering in the sphere of history cannot lead to positive results.[1]

I had no illusions that I had actually convinced Jarosław Kaczyński, but this exchange of opinions—if one can use this not entirely adequate phrase—foreshadowed what could happen if Law and Justice took power in Poland. Although it had less public resonance, Kaczyński's speech in September 2015 during the parliamentary election campaign was even more important, since the party was a frontrunner. At a meeting in Poznań he presented his vision for the reconstruction of the state. He declared the pursuit of a "decisive historical policy" that would carry an "affirmative message" to combat the "defamation of Poland" taking place abroad and would lead to the creation of "a front for strengthening the legitimacy of our state, strengthening the cohesion of our community." Museums would play an important role in achieving this goal. Kaczyński announced "the proper reconstruction of the Museum of the Second World War" (without explaining what that meant) and "changes to the Solidarity Museum"(European Solidarity Center in Gdańsk). He also proposed procuring, with Polish funding, a high-budget Hollywood film with a well-known director and star-studded cast, preferably about Captain Witold Pilecki, a hero of the Polish resistance movement. It was very characteristic, this conviction that an outstanding film, influencing the perception of Polish history in the world, could be produced on a political whim.

Kaczyński announced a change of cultural elites and promotion of new individuals in the domain of culture. He also criticized tendencies in contemporary culture that threatened traditional values and canons: "Society is provoked with all kinds of acts that in the West were seen or heard one hundred and more years ago. There is really nothing new in all these vulgar exploits."

As a historian who dealt with the Polish People's Republic governed by the Communists, I was reminded by the tone of some of these statements of a speech given by Władysław Gomułka at the famous ideological plenary of the Central Committee of the Polish United Workers' Party in July 1963. The first secretary

---

[1] Marek Adamkowicz, "PiS chce zmienić kształt Muzeum II Wojny Światowej. Jarosław Kaczyński kontra Paweł Machcewicz," *Dziennik Bałtycki*, July 1, 2013.

of the Communist Party subjected the sphere of culture and art to a detailed review, condemning specific creators and works that brought the "rot" of the West to Polish soil. Kaczyński's speech was very important; he presented a clear vision of the historical and cultural policies that he would implement after taking power. There was no room for institutional pluralism or museums that were not part of this unified front.²

We knew very well, therefore, the victorious party's feelings toward the Museum, and we soon learned how these sentiments would be reflected in specific steps taken against us and in the atmosphere created around us. We did not expect anything good, but the first blow was shocking, even to me, who personally knew many people from the "other side." These were my colleagues from studies at the University of Warsaw who had become Law and Justice politicians or journalists favoring the party; I had the opportunity to meet many others at the Institute of National Remembrance and got to know their mentality. Still, I was surprised when in November 2015, just after Law and Justice formed the government, Rafał Wnuk called me, upset. At a conference in Lublin, Piotr Niwiński, a colleague from our time at the Institute of National Remembrance, came to him, declaring, "We must talk about the Museum." "Paweł is a dead man," began Niwiński. "There is nothing more to say about him. Joachim [Brudziński] and Mariusz [Kamiński] told me that."³ He also said that he had seen, with his own eyes, "the list of people to be fired from the Museum." He announced that he might be able to save some of them, alluding to his close contacts with politicians from Law and Justice. In return, he expected cooperation from Rafał Wnuk and Janusz Marszalec, with whom he was ready to negotiate the appointment of a new Museum director acceptable to the party.

In the following months, he kept pestering Janusz Marszalec in Gdańsk, coming several times to the Museum when I was away. He repeated the message that I was a "dead man," claiming now that the person who had made that decision was Jarosław Kaczyński himself. He even maintained that Jarosław Kaczyński remembered me signing a 2010 petition to light candles on the graves of Red Army soldiers buried in Poland. This had been soon after the Smoleńsk catastrophe, in which president of Poland Lech Kaczyński and other Polish political leaders had died. Many sincere expressions of compassion had come from the Russian side at this time, and it seemed then that there was a chance for Pol-

---

2 Kaczyński's entire speech in Polish can be found online. See "Jarosław Kaczyński w AKO Poznań," video posted to YouTube by ako poznan on September 27, 2015, https://www.youtube.com/watch?v=E8Ggp24Y_G0. The Museum of the Second World War is mentioned at 0:37.
3 The latter was a minister supervising all secret services; the former was a minister of interior in the Law and Justice government.

ish-Russian reconciliation. I treated this not as a political but as a human and a Christian gesture. I was, of course, surprised that the party leader would take notice of me.

I did not know if Niwiński's statements were credible, since he certainly had no access to the party leader himself but rather based his opinions on what his friends from Law and Justice had told him. Undoubtedly, he was close to the ruling party; that was evident because he turned out to be one of three people asked by the Ministry of Culture to "review" our exhibitions. He was also appointed to a few museum councils, an example of the promotion of new people that Kaczyński had promised. At one point he admitted to Janusz Marszalec that he had written a review of our exhibitions for the minister of culture. Until it was released many months later, he maintained that it was positive and that he had suffered some displeasure as a result from the ministry. We quickly came to the conclusion that his objective was to replace me (since I was already a "dead man"). He even suggested that Jarosław Sellin, deputy minister in the Ministry of Culture, himself had urged him to take on the job. Apparently, he wanted my closest associates to ask him to replace me. "Help me choose a new director," he kept asking Janusz Marszalec when he came to the Museum in my absence.

A reader today may see this whole story as fairly grotesque. For us, however, it was the first, extremely unpleasant lesson about how someone we had thought was honest and knew well personally could behave. More broadly, it was a lesson about Poland after autumn 2015, when people began to follow the new order. Later on, we repeatedly found that no moral brakes would come into play, even in the case of existing friends and acquaintances, when it came to destroying the ideological enemy and taking over institutions and positions.

Immediately after the formation of the new government, Piotr Gliński and Sellin took over the Ministry of Culture, which initiated a series of hostile steps toward the Museum and myself. All our requests and applications, no matter what they pertained to, were rejected, generally without reason. For example, I asked Minister Gliński to use PLN 100,000 (€23,000) from our budget on the purchase of new servers, because the old servers were not meeting the needs of the growing institution. The answer was no, although I had not asked for additional money, only for a routine reclassification of the amount to the capital budget. That we were at risk of server failures, data loss, and setbacks in the huge investment of public money already spent on our institution did not matter. There were dozens of similar situations. Our daily work was made difficult in every possible way, even before the full-frontal assault of April 2016.

In December 2015, Sellin rejected my business plan for 2016. I routinely submitted a business plan every year, and this one was only a refinement and an update of the original plan, which I had prepared when I became the Museum's director. Sellin demanded to know "what experiences and what nations" would be presented in the Museum's main exhibition. On the one hand, this was a pretext to strike out against the Museum and myself. On the other, it revealed the attitude of Law and Justice politicians who reacted allergically to even the smallest indication that, in a museum devoted to the Second World War—a global phenomenon after all—experiences of any other nation than Poland might be shown.

I prepared an answer, which to me seemed absurd, since I had to explain why we did not limit ourselves to showing only Poles in the exhibitions. The following is an excerpt of my response that truly reflects the spirit of the times in Poland following the 2015 fall election:

> Due to the fact that the Museum has the ambition to write the Polish experience of the war into its comprehensive European and global dimension, it is obvious and necessary to show the actions taken by other countries and nations, both Polish allies, and aggressors and invaders: Germany and the Soviet Union. For example, we show in the main exhibition the campaigns in the West in 1940, including, of course, the Polish participation but also the French and the British. In the *Road to War* exhibition, we show the birth of Nazi and Soviet totalitarian regimes as the forces that overthrow the European order of the interwar period and plan aggression against Poland, which obviously requires showing the experience of the German people and the Soviet peoples. Similarly, the story of German and Soviet occupation and crimes committed against Polish citizens requires demonstration of the actions of representatives of the German nation and the nations of the USSR. The story of German and Soviet aggression against other states and nations as well as occupational systems created by both totalitarian powers constitutes part of the exhibition. Only against this background can one fully understand the specifics of the Polish wartime experience, both in Polish martyrdom and resistance.

Regardless of the somewhat grotesque nature of the situation and my explanations, this was a very serious matter. Government regulations gave the minister an opportunity to reject my plan for the next year, followed by removing me as the Museum's director. The amendment to the act on organizing and conducting cultural activities, introduced in 2012, actually strengthened the autonomy of directors of cultural institutions through multiyear appointments (from three to seven years) based on presented programs. This contract, in my case valid until the end of 2019, could only be cancelled if the director violated the law or failed to implement the program or the institution was liquidated. I was afraid that everything going on had to do with one of these three options, although I could not yet predict that it concerned the most drastic one, the formal liquidation of the museum.

In January 2016, after Sellin rejected my report, we knew that the end was imminent. From that moment, I lived, in a sense, in a state of suspension, prepared to leave the Museum and Gdańsk overnight. This was not about maintaining a position, which, with all its responsibilities, was primarily a burden and a source of constant stress that prevented a normal personal and family life. The most terrible prospect was that after almost eight years of work, when we were very close to the end and to opening the Museum, all this could be interrupted, and we would not even be able to show the results of our activities to the public.

Perhaps the worst part of this whole situation, which lasted over a year, was the total uncertainty, not knowing what might happen the next day, what measures might be taken against the Museum and me. Despite all this, it was necessary to supervise construction that was entering the final phase, to solve dozens of problems that arose every day, resolve disputes with the general contractor and the exhibition producer Tempora, and decide on spending tens of millions of zlotys. This was done in an extremely unfavorable external environment. We were fully aware that the Ministry of Culture would use every mistake or risky choice against us and, at the same time, would do everything possible to make life difficult for us, from construction to even minor matters. Avoiding decisions would have protected me and the others responsible, but it would have delayed the work. And our main goal was to present a fait accompli, something difficult to reverse. First of all, we wanted to begin installation of the exhibitions in the museum building, which was to start in the second half of the year. We knew that changes to even a partially finished exhibition would be much more difficult; they would also cause an enormous scandal and entail huge financial costs. In fact, we did not believe that we would be able to open the Museum; we still needed a lot of time. We wanted, however, to do as much as possible while we could. I only told my three closest associates, Janusz Marszalec, Piotr M. Majewski, and Rafał Wnuk, that we probably would not succeed.

We tried to mobilize the team, by then comprising several dozen employees, showing at every step of the way that we were not giving up, not surrendering, giving more than ever before, and expected the same from the others. We picked up the pace of work everywhere, although the construction could not have been performed at a faster pace. So it was not only a constant struggle with the Ministry of Culture to let us finish the Museum but also a race against time, of which we did not have much left. We hoped that the chaos of the new government in the first months after winning power would be our ally, but we were well aware that the frontal assault would come sooner or later.

Despite all the bad signs, I did not completely lose hope that a substantive conversation could take place about the merits of the exhibitions. During the election campaign in 2015, Jarosław Sellin, who did not hide his ambition to become minister of culture, announced that the exhibitions of the Museum must be changed so that they would become more "Polish." This reflected the party line of Law and Justice, so we were not particularly surprised. He also asked us to provide materials on the exhibitions, with which he wanted to become acquainted personally. We did this at the beginning of January 2016, and when I met Sellin at a public meeting on historical policy with Polish president Andrzej Duda, he told me that he was studying the material and was going to come to Gdańsk to meet with us and talk about the shape of the exhibitions. He even gave a specific time frame, the second half of March. I knew Sellin from earlier times; we were even on a first name basis. My wife and I used to be close friends with Joanna Lichocka, a journalist and later a Law and Justice MP, and she had introduced us to Sellin, who at the time was an opposition MP. So we waited for the visit of Sellin, the secretary of state responsible for museums and historical policy, but it never materialized.

Minister Gliński Comes at Night

It was Friday, April 15, 2016, and as usual I had returned from Gdańsk to Warsaw after a week of work. A few minutes after 11 p.m., just before turning off the lights and going to bed, I noticed a text message from Janusz Marszalec asking me to call him immediately. I had been expecting the worst for many months, but what he told me still came as a complete surprise. A friend of Marszalec had happened upon an announcement that appeared that evening on the website of the Ministry of Culture. The minister of culture and national heritage had announced the liquidation of the Museum of the Second World War. This was to be done through a merger with the Museum of Westerplatte and the War of 1939 to create a new cultural institution bearing the name of the latter. Here is how this step was justified:

> The reason for the merger is to ensure increased effectiveness of measures to: maintain and promote national and state traditions, in particular in reference to the history of Poland during the Second World War, including the defensive war of 1939; proper commemoration of the symbolic place which is the battlefield of Westerplatte; and coherent implementation of supervision over museums for which the minister of culture and national heritage is the organizer through optimizing the potential of branches with a similar business profile located in Gdańsk, while rationalizing the spending of funds from the state budget.[1]

The very name of the Museum of the Second World War would disappear. The institution would be absorbed by another museum, which instead of a global conflict would deal only with the seven-day defense of the Military Transit Depot at Westerplatte and the war of 1939. Both events were widely presented as part of our exhibitions, but 90 percent of the content would not fit into these very narrow parameters. What about the rest? Would it mean dumping eight years of work by dozens of historians and designers and devoting this huge, almost finished building only to one month of the war, which in the context of the Polish fate between 1939 and 1945, although important, was just a fragment? In any case, it was a plan for the total annihilation of the hated museum, whose name would not even survive.

---

1 "Obwieszczenie Ministra Kultury i Dziedzictwa Narodowego z dnia 15 kwietnia 2016 roku o zamiarze i przyczynach połączenia państwowych instytucji kultury Muzeum II Wojny Światowej w Gdańsku oraz Muzeum Westerplatte i Wojny 1939," *Dziennik urzędowy Ministerstwa Kultury i Dziedzictwa Narodowego* 18 (April 15, 2016), http://bip.mkidn.gov.pl/media/dziennik_urzedowy/p_18_2016.pdf.

Regardless of the nature of the announcement, both the circumstances and justification were shocking. The decision to liquidate a large museum, after eight years of work and shortly before the anticipated opening, was not communicated to its director, who was responsible for a project worth several hundred million zlotys, the largest cultural investment in Poland at that time. Everything was done by surprise, and the release of information on the ministry's website just before the weekend, on Friday afternoon or evening (I was unable to confirm a specific time), was intended to minimize protest. As I learned later from the people I knew at the Ministry of Culture, it was hoped that by Monday new, more important topics would attract the media's attention. Even more astounding was the justification presented by the ministry: "optimal use of the potential of branches with a similar business profile, located in Gdańsk."

The Museum of Westerplatte and the War of 1939 was established on December 22, 2015. I was a bit surprised because in our exhibitions we devoted a lot of space to these themes. Above all, at Westerplatte, as I have already mentioned, a branch of the Historical Museum of the City of Gdańsk was operating at Guard House No. 1. For many years, there were also two permanent outdoor exhibitions produced by our Museum, annually visited by tens of thousands of people. There was no merit in the establishment of a new museum at Westerplatte, in my opinion. I thought that deputy minister Sellin simply wanted to have it his way and was returning to his original proposal brought up during the first term of the Law and Justice Party. He treated the whole affair as a personal matter. For many months, the new museum showed no activity, but we had so many things on our minds that we did not think about it. Now it was clear that from the very beginning it had been a Trojan horse, a well-conceived plan to use a museum that existed only on paper to liquidate the Museum of the Second World War, end my tenure as director before my contract was up, and prevent us from presenting to the public the exhibitions on which we had been working for all these previous years.

After April 15, dozens of journalists began to come to Gdańsk to talk to me; they also tried to reach Mariusz Wójtowicz, director of the Westerplatte Museum. That proved impossible because the museum did not have an address, a website, or a telephone. It was a "ghost museum" of which nothing was known except that it was to absorb the Museum of the Second World War. Even in March 2016 the Westerplatte Museum had no staff, and it only began to hire its first employees in the next month, when there was lots of noise around the "ghost museum" and some steps had to be taken to avoid public backlash.[2] The cynicism of

---

2 An answer of the director of the Museum of Westerplatte and the War of 1939, Mariusz Wój-

the repeated public assurances by Minister of Culture Piotr Gliński and his deputy Jarosław Sellin—that their decision was about saving public funds and combining the potential of two institutions operating in one city—was striking. The potential of the fictitious museum was not at all clear. Nobody believed the specious argument, but with a poker face, the Law and Justice politicians and their allies used it to justify their actions. "Two ... museums are not needed in the same city. It costs Poland very deeply," said Minister Gliński, who himself had created the second museum a few months earlier.[3]

A separate issue was the person placed at the head of the Westerplatte Museum by Sellin, who treated Pomerania as his backyard and certainly made all the more important decisions that affected the region. The life mission of Mariusz Wójtowicz, an amateur historian and an engineer by training, was to contest the legend of Major Henryk Sucharski, commander of the Military Transit Depot. He accused him of cowardice because Sucharski had wanted to surrender Westerplatte as early as September 2, after a devastating airstrike by German bombers and pursuant to orders that required him to defend the isolated facility for up to twelve hours. Wójtowicz insinuated that Sucharski was a demoralized man, sick with syphilis, a homosexual with a "clear weakness" for his own orderly, and even a German agent.[4] Wójtowicz's life's work, the book *Westerplatte: The Real Story*, was crushed by professional historians who proved that he had manipulated his sources and displayed obsessive hatred toward Sucharski.[5] Now such a man was at the head of the newly created museum that was supposed to "swallow" the Museum of the Second World War. Moreover, he was appointed to this position by the government and the party that constantly declared the need to "restore pride in our history." Westerplatte was one of the most indicative symbols of this ambition.

After Janusz Marszalec read the text of Minister Gliński's announcement to me that night, I was convinced that this was the end and that there would not be a trace left of our eight years of work on the exhibitions. At first, I thought that the Museum was being liquidated immediately, and I even wondered how I would retrieve my personal belongings. I thought that on Monday I would no

---

towicz, to the question posed by Wojciech Ogrodnik, submitted via the Access to Information legislation, October 4, 2016.
3 "Gliński: Dwa muzea w tym samym mieście są bardzo kosztowne," April 18, 2016, Polsatnews.pl, April 18, 2016, https://www.polsatnews.pl/wiadomosc/2016-04-18/glinski-dwa-muzea-w-tym-samym-miescie-sa-bardzo-kosztowne.
4 Mariusz Wójtowicz, "Major Sucharski—zdrajca i tchórz?," *Odkrywca* 9 (2007).
5 See, among others, Jan Szkudliński, "Westerplatte 1939. Prawdziwa historia?," *Dziennik Bałtycki*, September 29, 2009.

longer be allowed in. Nevertheless, I tried to do everything to ensure we were not "liquidated" quietly. On Saturday morning I sent an email to over a hundred people—historians, museum workers, journalists—with information about what had happened and attached the text of the minister's notice, downloaded from the ministry's website. I did not expect that this would start an avalanche, which would grow throughout the following year, extending the lifeline to the Museum and, above all, giving us a chance to open it.

The reaction was immediate. Over the weekend I gave many interviews to newspapers and radio stations, and information about the Museum's situation found its way to the forefront of the news services. The mayor of Gdańsk, Paweł Adamowicz, called a press conference. He announced that the city had donated the land for the construction of the Museum of the Second World War and that the deed stipulated the gift could be withdrawn if the museum was not created. As it turned out, this decisive argument saved the name of the Museum of the Second World War. The Ministry of Culture was completely taken by surprise, not having known that such a clause existed. On Monday I was asked to send over the deed. I am convinced that this was the main reason that Gliński and Sellin modified the plan for the total annihilation of the Museum of the Second World War. On May 6, the minister of culture made a new announcement. He upheld the intention to merge the museums; he literally repeated the argument from the announcement of April 15. The new institution, however, would be called the Museum of the Second World War. This did not change the fact that both institutions would be formally closed down. A "new" Museum of the Second World War would be created. My contract would be cancelled. I would not even have to be dismissed from my job; there would simply be no more museum of which I was the director. Most importantly, this would still open the way to the complete transformation of the main exhibitions.

Equally important were the voices of protest and support for the Museum coming from historians. During the first weekend alone, two professors from the Institute of History of the Polish Academy of Sciences, Maciej Janowski and Adam Manikowski, collected hundreds of signatures under an open letter. Among the signatories were the most prominent Polish historians, not only people known for their opposition to the Law and Justice government but also those who, I knew, sympathized with it, or in any event were far from those who had held power in Poland before the election in 2015. I will quote only the last portion of the open letter: "It is difficult to avoid the impression that the decision to liquidate is part of the logic of the political struggle, consisting of the destruction of institutions established by the previous government, completely independently of their substantive values. As historians, humanists and museum professionals, it is difficult for us to accept acts of vandalism thoughtlessly carried out on our culture."[6]

Two weeks later, 198 American and European historians sent an open letter to the minister of culture. Among them were the greatest names in our discipline from several dozen universities representing various ideological environments (for example, in the United States both conservative Republicans and people with liberal or leftist views)—in other words, people who do not usually sign the same letters. They stated that to prevent the opening of nearly completed exhibitions after eight years of work would be "an unprecedented injustice in the democratic Western world."[7]

The veterans' communities also protested. Just a few days after the minister's announcement, the Council of Veterans and Repressed Persons, representing all such organizations in Pomerania, expressed its opinion:

> The Museum of the Second World War was supposed to be a showcase of Poland in Europe; it had a chance to join the sensible European historical politics and show our country in the context of the anti-Hitler coalition... . The Museum of the Second World War has a largely constructed building, a long-established master plan; has acquired thousands of artifacts, which were also collected by our members, the Second World War veterans... . We are appalled that the government, for which historical policy is so important, wants to liquidate the museum that presents the Polish perspective on the Second World War.[8]

The National Union of Soldiers of the Home Army, the most important veterans' organization in Poland, also defended the Museum. The words of support from veterans were of special importance to us. It was difficult to accuse them—unlike me and other authors of the exhibitions—of a lack of patriotism; yet they were consistently dismissed by the minister of culture. They received evasive answers to their letters. When they asked for a meeting to present their arguments in defense of the Museum, the minister delayed for many months and tried to pass their request down to a lower level of the ministerial bureaucracy. For me, it was particularly important that veterans supported the comprehensive context of the exhibitions not limited to the Polish experience. They had much broader horizons and, better than the autocrats of the Ministry of Culture and apparatchiks of the ruling party, understood that the history of Poland was best explained against the background of the fates of other nations.

---

6 "Protestujemy przeciw likwidacji Muzeum II Wojny Światowej," *Gazeta Wyborcza*, April 18, 2016.
7 "List otwarty do Ministra Kultury Piotra Glińskiego," *Gazeta Wyborcza*, May 5, 2016.
8 Oświadczenie Rady Kombatantów i Osób Represjonowanych Województwa Pomorskiego, April 21, 2016.

Many donors, thanks to whom we had been able to build the collection and create the exhibitions, also declared concern and solidarity with us. After the minister's announcement about "joining the museums," several dozen people contacted us and stated that they would consider withdrawing the artifacts handed over to the Museum. These were some of our most valuable objects, constituting the core of the main exhibition: for example, the already mentioned memorabilia of Józef Stach, shot in Piaśnica; personal objects of Colonel Aleksander Wilk-Krzyżanowski; and the banner of the 6th Heavy Artillery Regiment from Lviv, hidden after the Soviet invasion.

We asked donors not to make hasty decisions to withdraw donations, but we often met with very emotional reactions that were also expressed in conversations with journalists. Andrzej Stachecki, who gave the museum artifacts from his father murdered in Piaśnica and father-in-law imprisoned in Stutthof, asked, "What do Misters Gliński and Sellin know about these things? ... They are playing politics at the expense of someone's suffering, someone's drama, someone's life! I do not agree that the paws of politicians dig in these relics. I gave these items to a specific museum, to people I trust. If they are to be fired and the museum is to become something else, then I take these mementos back. They will be better off in my son's home."

Stachecki also referred to the accusations that the Museum did not represent the Polish point of view:

> My father was a Pole, the greatest patriot... . [H]e gave his life for Poland! So how can they say that? Let them learn the history of Poland! This is a pity for my father and his memory! This is a very Polish museum, a kind of monument for ordinary people who gave their lives for this country. Not for leaders, not for politicians. My mother hid my father's things from the Gestapo, then from the UB [Communist security services], and I will take them away from the museum so that politicians will not manipulate them.[9]

These emotions proved that a museum and its exhibitions are not only an institution or intellectual work created by historians. They are also a record of very authentic feelings of people who passed on to us part of their family history and who treat exhibitions as tributes to their loved ones and, in a sense, as their own creations. It was fascinating and moving for us to observe how a social movement was being formed to defend the Museum. We were not alone.

The Museum was also defended by unknown people who had had nothing to do with it before. Sociologist Wojciech Ogrodnik opened a Facebook page ti-

---

9 Sebastian Łupak, "Darczyńcy Muzeum II WŚ: wycofamy nasze pamiątki," Gdańsk.pl, April 24, 2016.

tled "No to the Liquidation of the Museum," where he published all the information and collected signatures for further and further appeals in our defense. Thousands of Internet users visited it, and it became an important communication channel for people who supported us and even a kind of platform through which new initiatives were formed. Activists of the Committee for the Defense of Democracy, from not only Pomerania but also other, even more distant regions of Poland, organized pickets in front of the Museum building, during which they hung white and red ribbons (Polish national colors) with messages: "History is nonpartisan." "Stop falsification of history." "We support the Museum of the Second World War." "We are with you." "Leave our museum alone." "History is not yours; it belongs to all." "There is one History!!!" "Second World War, not only Polish." "Gliński, hands off the Museum." "We do not agree to this shameless change." "War is a drama and tragedy for every man." The ribbons were removed by "unknown assailants," but new ones always appeared. Lots of people came around to look. I saw them reading messages and photographing ribbons.

I was not the only one feeling that something extraordinary was happening. No museum in Poland has ever engendered such emotion and enticed ordinary people to spontaneously express solidarity and support for the endangered values with which they identified. To me and my closest associates, it was now an obligation. Now, it was not just about ensuring that something remained of our many years of work; we could not disappoint all those who were on our side.

The Race Against Time

With the media storm, the public outrage, and certainly the announcement of Gdańsk mayor Paweł Adamowicz that he would take back the donated land, Minister of Culture Piotr Gliński clearly realized that the Museum could not be dealt with silently and without a loss of reputation to the "liquidators." On Sunday, April 16, I received a call from the Ministry of Culture with an invitation from Minister Gliński to meet the next day. The conversation in his office lasted an hour. The minister was ostentatiously polite, apologizing to me for the "nonstandard" mode of operation—for example, not informing me about the decision to close the Museum that I managed. Nevertheless, the conversation was full of surreal elements. The minister treated me as a representative of the ancien régime and presented me with all the charges against the previous authorities, beginning with the fact that the Museum of Polish History had still not been created. I explained that such allegations should be addressed to former prime ministers and ministers of culture; I was responsible for creating the Museum of the Second World War, which I was ready to discuss. I also vigorously explained to the minister that restricting exhibitions to only the events of 1939—the logical consequence of his own decision to liquidate our Museum—did not make any sense; nor did it correspond to the goals declared by his government. The whole experience was bitterly absurd because I had to explain obvious issues with a straight face: the inevitable omissions in the new museum that would result from the merger, including themes of fundamental importance to the entire Polish wartime experience such as the terror of both occupiers (including at Auschwitz and Katyń), the Polish Underground State, Operation Tempest (a series of anti-German military operations in 1944), the Warsaw Uprising, and the Holocaust of Polish Jews. I also explained the real nature of our exhibitions, noting that the vast majority of attacks on them were formulated by people who had no idea, and did not want to have any idea, of the content of the exhibitions. I took notes on the conversations, which I placed on the Museum's website the next day: "In response, Minister Gliński stated that if this was the nature of the Museum's exhibitions, this would create a good platform for further discussions."

Gliński formulated only one specific complaint against the exhibitions: that they allegedly did not include crimes committed by Ukrainian nationalists on Poles in Volhynia. I explained that it was just the opposite: we were the first museum in Poland and probably in the world to have a separate section devoted to the extermination of Poles in Volhynia in 1943 and 1944, which contained, among other things, objects belonging to victims and gathered during exhumations. Apparently, the minister did not bother to make up his own mind about the exhibitions but relied on commissioned reviews. We later learned that the author of one of these, Piotr Semka, had complained that we did not show the crime in Volhynia, commenting, "It's a real scandal!" He had overlooked the information included in the overview of the exhibitions, which was among the materials we had previously submitted to the Ministry of Culture.

During our meeting Gliński talked about four critical reviews and proclaimed he would disclose their contents and the names of the authors and organize a debate between the reviewers and the creators of the exhibitions. He later withdrew this proposal, but on April 18 it seemed that there was a chance for a dialogue that would allow us to present our reasons and, above all, show the actual shape of the exhibitions. Gliński stressed that the announcement about the liquidation of the Museum was not final; he gave himself at least three months to make a decision about the future of the Museum of the Second World War. Going into this conversation, I already knew that an act pertaining to organizing and conducting cultural activities required passage of at least ninety days from announcement of the intention to close or merge cultural institutions to actual implementation of these plans. Theoretically, this time was to be devoted to public consultations. The situation, therefore, was not as hopeless as it had seemed a few days before when, late at night, I found out about Minister Gliński's announcement. We had at least three months to fight for the Museum's survival or to leave, in some way, a tangible trace of our work.

It was obvious to me that given the immense difference in forces, reminiscent of the clash between David and Goliath, informing the public was essential so that we would not be alone. That was not really difficult because from the moment the government announced its plan to liquidate the Museum and throughout the year that followed, the matter received huge and unflagging media attention. I barely managed to get out of Gliński's office before I was taking calls from journalists, and I immediately reported the course of the conversation to several radio stations. The intention to close a huge museum that had been under development for eight years and was about to open (and the announcement of this decision in a brief message on the ministry's website) seemed so brutal and unprecedented that it exceeded even the low standard set by Law and Justice in the first months of its rule.

We did not need to stir up journalistic interest and public outrage; they arose spontaneously and naturally. This was a real thorn in the side of Gliński and Deputy Minister Jarosław Sellin, who apparently had hoped we would be "strangled in silence," with moderate protests quieting down after a few weeks. When this proved not to be the case, they began to accuse me of inciting public opinion in the country and abroad and even conducting an organized campaign to discredit the Polish government; Gliński threw this very allegation at me a few months later at a meeting of the Senate's cultural committee. They probably believed it sincerely; Sellin confronted me with the same charge during a backroom conversation at a meeting of the International Auschwitz Council. This clearly reflected their conspiratorial vision of the world as well as their distrust of the elementary mechanisms of democracy, such as a free media; these are not so easy to control, especially not by the director of a museum in Gdańsk.

These kinds of accusations were yet another grotesque feature of this whole, rather gloomy story. I mocked them, stating that they flattered me but at the same time overestimated my ability to inspire the world's largest newspapers and TV stations, which were reporting extensively on the Museum. The *New York Times*, for example, made it one of the main stories of the paper's cultural section. Vanessa Gera, Warsaw correspondent for the Associated Press and a historian by training, devoted many of her stories to the Museum, and these were reprinted or quoted by dozens of newspapers and internet portals around the world.

I also had no influence over the public speeches of such internationally renowned historians as Timothy Snyder or Norman Davies. The former published a very widely read article in the *New York Review of Books*. Later, in the United States and Europe, I often met people who had read and remembered this article. Snyder emphasized the uniqueness of our Museum, which, according to him, had the most ambitious character among all world institutions dealing with the Second World War. The author of *Bloodlands* wrote, "Unlike other museums devoted to history's most devastating war, which tend to begin and end with national history, the Gdańsk museum has set out to show the perspectives of societies around the world, through a sprawling collection gathered over the last eight years, and through themes that bring seemingly disparate experiences together. It is hard to think of a more fitting place for such a museum than Poland, whose citizens experienced the worst of the war."

He explained that the Museum would show not only the war but also the path to it—the birth and expansion of totalitarianism in the 1920s and 1930s—and would offer an innovative interpretation of humanity's most important experience in the twentieth century. Inclusion of the Holocaust in a wider landscape of German crimes provided a better opportunity to understand it than did museums dealing only with the Shoah. Similarly innovative, according to Snyder, was the cross-sectional, transnational display of such issues as collaboration, the fates of prisoners of war, and the bombing of cities. In other museums these were usually shown only in the context of their own national experiences, which inevitably impoverished the perspective and did not allow visitors to truly understand the nature of events. Snyder devoted a large part of his article to the argument that turning an almost-ready "global" museum into an obscure institution, which would deal only with the seven-day defense of Westerplatte, was detrimental to Polish history. Hence, the dramatic title of his article: "Poland Versus History." The first word in the title, of course, referred to the Polish authorities. Snyder listed topics central to the wartime experience of Poland that could not be found in the new "substitute" museum: the terror of the invaders, the Holocaust, resistance in the country, and the struggle of Polish soldiers on all fronts. He gave examples of artifacts that would not be shown in the Westerplatte Museum: the handkerchief of Bolesław Wnuk, murdered by the Germans in 1940; the Enigma cipher machine; keys from Jedwabne. He asked whether preventing the display of controversial artifacts might have been, in fact, one of the objectives of canceling the museum.

It reminded me a bit of my own conversation with the minister of culture, in which I tried to convince him that the cancelation of our Museum would deprive Poles and the whole world of the opportunity to see what was most important in our history. Snyder, however, added new, important reasons, noting that it would be very handy for Vladimir Putin.

> Perhaps the greatest surprise in the Polish government's decision is the implicit alliance with current Russian memory policy. The move to limit the Polish history of World War II to the week-long engagement with Germany at Westerplatte in 1939 follows a Russian script... . And the Gdańsk museum has collected the stars from the uniforms of some of the 22,000 Polish officers murdered by the NKVD in the Katyń massacre in April 1940, a humble reliquary of those Soviet death pits. Once the museum is out of the way, the Kremlin can be confident that no one else in Europe (beyond the Baltic States) will make the attempt to inscribe the Soviet aggression of 1939 and the occupation regime of 1939–1941 within the public history of the war.

Snyder finished his text with a prophetic warning: "In some measure at least, how rising generations of Poles see themselves, democracy, and Europe will depend on whether they can have ready access to their country's complicated experience in World War II. The collapse of democracy, the museum's first theme, could hardly be more salient than it is right now. And the presentation of the conflict as a global tragedy could hardly be more instructive. The pre-emptive liquidation of the museum is nothing less than a violent blow to the world's cultural heritage."[10]

I do not know if Snyder's text, reprinted and discussed also in the Polish media, influenced the thinking of Minister Gliński, but a few days later he an-

---

10 Timothy Snyder, "Poland vs History," *New York Review of Books*, May 3, 2016, https://www.nybooks.com/daily/2016/05/03/poland-vs-history-museum-gdansk.

nounced—as I have already mentioned—that the Museum of the Second World War would survive, though as a new cultural institution (and probably with other exhibitions). However, I am inclined to think that the decisive factor was the mayor of Gdańsk's announcement that he would reclaim the donated land if the museum simply ceased to exist. We knew for sure only that we still had three months ahead of us, and we wanted to create as many faits accomplis as possible that would later be difficult to reverse. Fortunately, building construction and exhibition production could not be stopped because I had managed to sign all the most important contracts, and breaking them would result in millions in penalties for the treasury and an unimaginable scandal that even the Law and Justice government could not afford.

Of course, our priority was completing the exhibitions and having the opportunity to show them to the public. Production began at the end of 2015. We initially planned for the installation to take more than ten months, which was still a very ambitious schedule in the case of such a large museum with such a complex design. After discussions with the exhibition's fabricator, the time frame was shortened to five months, although it was not at all clear whether it would be possible to stick to these tight deadlines. This required perfect coordination with the final phase of the building construction.

We also decided to accelerate work on the catalog for the main exhibition, which we started to prepare immediately after the clouds began to gather over the Museum. A catalog is almost always made available to the public only with the opening of a museum. We considered the possibility that the opening might never happen and that the catalog might become the only record of our eight years of work on the exhibitions. It was, on the one hand, something like our testament, as we said in a fit of black humor (then a fairly frequent state of our spirits). Considering that, according to Piotr Niwiński, I had been a "dead man" for several months, this analogy was apt. On the other hand, the catalog was of fundamental importance for us to inform the public about the actual shape of the exhibitions and to refute all the lies. In June 2016 we managed to publish the catalog, first in the Polish version and a few weeks later in English.[11] We sent it to a wide group of historians, museum professionals, and journalists, and we also offered it for sale online. We had the satisfaction that, after all our work, at least this would remain: a record of a nearly finished exhibition that might never open.

---

11 Rafał Wnuk et al., *Muzeum II Wojny Światowej. Katalog wystawy głównej* (Gdańsk: Muzeum II Wojny Światowej, 2016).

Parliamentary Committee on Culture: Hunting for "Encyclopedists"

If my conversation with Minister Gliński raised my expectations for a dialogue about the Museum and the shape of the exhibitions, the session of the Sejm's[12] Culture Committee in June 2016 stripped me of all illusions. It was convened at the request of the opposition deputies who wanted information from the minister of culture about his plans for the Museum of the Second World War, but it turned into a public judgment on the Museum pronounced by the members of Law and Justice.

Minister Gliński began by presenting known arguments for the merger: "Of course, the main reason was the fact that there was no point in having two state institutions in one city that covered the same scope." Opposition MPs reminded him that in December 2015 they had asked why it was necessary to set up another museum in Gdańsk dealing with the war. At the time, Deputy Minister Sellin had assured them that there would be a place for both the Museum of the Second Word War and the newly created Museum of Westerplatte and that he saw no problem with creating a second institution. Minister of Culture Gliński did not address the issue; he went on instead about the alleged "neglect" of Westerplatte and stigmatized the previous government's failure to build the Museum of Polish History. He also admitted, quite honestly, why he could not take even more radical steps toward our Museum: "The facts that took place, in other words the practical spending of the sums earmarked for this investment, that is, the facts and contracts made, would be a waste of public funds, if we tried to stop the project being carried out at the Museum of the Second World War."

Gliński's speech was quite moderate compared to those of Law and Justice deputies, who spoke in turns. Dariusz Piontkowski accused us of representing "a cosmopolitan historical narrative of the world." The MP from Białystok (previously marshal of the Podlasie region) elaborated:

> Here the main dispute takes place. Is the museum supposed to be a cosmopolitan museum where British and American historians will tell us how we, Poles, should imagine the Second World War and how to portray it, or, as we argue, should we show the world how we see the Second World War and how it influenced the fate of Poles? This is a fundamental dispute. From our point of view, the exhibition prepared by the director, unfortunately, rather corresponds to the demand of this cosmopolitan vision of history, which is partly detached from the needs of Poles.

---

12 The Sejm is the lower house of the Polish parliament.

He also wrote our Museum into the "anti-Polish" historical policy of the previous government:

> We, Law and Justice, before the elections clearly stated that one of the areas in which we did not agree with the ruling coalition was, among other things, the vision of historical policy—or lack of vision, because sometimes we had the impression that this vision did not exist. We have made it clear that many figures in political and intellectual milieus—unfortunately, previous governments have succumbed to these environments—have tried to conduct a pedagogy of shame, concealing, as much as possible, the Polish perspective on the past, not only on the Second World War but also on more distant times, and exposing these points of view that from the point of view of Poles were not only neutral but even hostile to Poland and Polish tradition.

Arkadiusz Czartoryski, also a historian by education and at one point a history teacher, accused us of omitting in our exhibitions the stories of the Monte Cassino battle, the 303 Squadron, the extermination of Polish intelligentsia and priests in Pomerania, Auschwitz, and Dachau, and the rescue of Jews by Żegota (the underground Council to Aid Jews) and other Poles (of course, all these subjects were, in fact, presented). He found in the "Outline of the Exhibitions"—the same document used by the reviewers of the Ministry of Culture—artifacts that roused his objections and demanded their removal. These were the pipe of Stalin ("a donor ... can take this pipe back") and a tapestry embroidered in the prison in Gdańsk by the SS female guards from Stutthof concentration camp before their execution in 1946. Given to the head of the prison as an expression of gratitude for their proper treatment, this was one of the most unusual artifacts in our collection, obtained from the governor of a prison. For the deputy from Ostrołęka, however, this was a tool in our unfortunate attempts to distort historical truth: "If you talk about the general narrative in the context of the whole of Europe, I wonder. A young man from Germany, for example an eighteen-year-old, will come to the Museum of the Second World War, see the tapestry, a beautiful white eagle, and read the caption: a tapestry with an eagle embroidered by German guards of the Stutthof camp before their execution. Some poor German guards embroidered such a beautiful tapestry, and were later shot, killed."

In conclusion, Czartoryski stated that he opposed this type of exhibition:

> I, as a person who in the Sejm takes part in the adoption of the budget, when hundreds of millions of zlotys paid by Polish taxpayers are allocated for various expenses, I do not agree that the first museum, which is built for half a billion, above all, presents things that are secondary from the point of view of the essence of the Second World War in the context of how we Poles see it. I would like the world to learn about how Polish couriers crisscrossed Europe, about how we saved Jews, and about the genesis of the murder of the Polish nation. Please note that in Berlin there is a monument to the murdered Jews, the

murdered Roma, the murdered Sinti, but there is no monument to the murdered Poles anymore.

The written record of the Culture Committee meeting does not reflect the real atmosphere and the temperature of emotions. For me, it felt like a public lynching. MP Czartoryski, near whom I sat, shouted, flushed, and gesticulated. He and other colleagues from his party really seemed to believe that they were dealing with enemies of their homeland, whom they had managed to stop at the last moment and prevented from opening an "anti-Polish" exhibition. Joanna Lichocka also responded, although she had to overcome her personal scruples. She noted, "I have known Professor Paweł Machcewicz for many years personally; I really like him and also value him as a historian." The truth, however, was more important:

> At the moment, the project of the Museum of the Second World War is a project showing the course of the Second World War from the perspective of the martyrdom of civilians, and not, for example, showing war achievements like the military struggles of the Polish Underground State or other events of this type that my colleagues have already mentioned. This view of the Second World War as a martyrdom of civilians is very convenient and very close to the German concept of historical politics. This is my opinion, but it is also an opinion of, for example, Piotr Semka, who in recent years has been publicly speaking very critically about this subject. So, I would like to ask, Why does Westerplatte bother you? Is it because the heroism of Polish soldiers in a collision with the German invader is such a strong symbol that it will turn upside down the ideas adopted by the [opposition party], which are so close to German historical policy?

The discourse was not without odd and even grotesque threads, although during the committee's hearings, I perceived them rather differently, not appreciating their comicality. Anna Sobecka, MP from Toruń, linked to the Catholic-nationalistic Radio Maryja, expressed concern about how we treated one of the artifacts:

> Director, the Sherman tank is one of the most valuable artifacts of the Museum of the Second World War, of which you are probably very proud. As far as I know, this is a donation from the Belgian government. The Museums Act obliges you, as the director of this investment, to care for artifacts and collections. Why, then, was one of the few operating vehicles of this type in Europe sunk in concrete? Is the engine removed from the vehicle? I hope I will get an answer, if it is not true. I have followed the reports on this subject well, and I say what I know. Will the tank be maintained, and what will happen to the artifact when there is a sudden roof leak?

This was a continuation of attacks on the Museum orchestrated by Sellin, who had previously accused us in the Sejm of sinking the tank in concrete. According to him, instead of making it part of the exhibition, we should drive it around Poland and promote Polish history. It was a good example of how every possible fact associated with the Museum was distorted to attack us. In turn, MP Jacek Świat alleged that the Museum building was deliberately hidden underground in collusion with the developer who built the apartment buildings on the neighboring lot. In Gdańsk "it is said that a large building constructed not underground, but above the ground, a museum building, would simply hide the view of the development complex." I was also accused of building the museum on "wet ground" (it is interesting that a housing estate was being built right next to it) and of "extraordinary extravagance"; several deputies appeared as experts on the costs of archeological research, accusing me of repeatedly overstating them.

Elżbieta Kruk, the head of the Culture Committee, did not allow the invited representatives of the Museum's Trust Council to speak. My answers were limited to a minimum, and I was not permitted to respond to all the allegations. However, she did give the floor to and allowed very long speeches from people who neither sat on the committee nor had any other authority to speak. Jan Żaryn, who was not a Sejm deputy, accused the creators of the exhibitions of lacking Polish national sensitivity and of "universalism." He also declared that in the approach to history, one should be inspired by the Jews, although he could not bring himself to say that last word:

> You all know perfectly well whom we can and should follow. This is also clear in this program. The only exception that has been introduced to this universal story, the right exception, to be clear, is the history of the Holocaust. Nobody is able to, or brave enough, of course—if you must treat this in the category of courage—to exclude the Holocaust from the history of the Second World War and consider it as one of many, serving universal values. You did not do it either, Director. And justly. But why should Poles, for the first time as an independent nation and state, not do the exact same thing, so that later when the Museum of the Second World War is built, no one will dare to push us into universalism, only in experience as strong as was the Holocaust. Who else should do this, if not Poles?

A TV journalist, Jan Pospieszalski, sat in one row with the Law and Justice MPs. I do not know in what capacity he was present, but the chairwoman of the committee gave him the floor and let him speak for a good ten minutes. He included our Museum in a landscape of historical revisionism and accusations against Poles: "A few days ago, at a book fair, the largest book fair in Chicago, there was a significant event. Just when a lot of people had gathered from around the world, a picket of Jewish communities accused Poland of fostering extermination, of bringing about the Holocaust, of participation in the extermination of Jews." Similar observations led him to an unequivocal conclusion, seasoned with additional mockery: "The museum should be created exactly with this scenario, but in Brussels. Perhaps there, there will not be such wet areas."

As the meeting progressed, I had an increasingly strong sense of a growing nightmare. I knew that the allegations had no relation to the actual shape of the exhibitions, our work, and our intentions. They were simply part of a public spectacle that was meant to stigmatize the enemy, trample him into the ground. Elżbieta Kruk, chairwoman of the committee, and all of the Law and Justice MPs who spoke were prepared for this task, and the course of the entire committee hearing was predirected in advance. Today when I read the final statement I gave, I am embarrassed by the fact that I was explaining my "Polishness." I emphasized that the exhibition presented Polish historical experience and that for most of my life I had dealt with Polish history. I will quote the final part of the committee's hearing, which reflects its atmosphere and my desperation to explain that we were not guilty of national dishonor and disgrace or "cosmopolitanism."

Director of the Museum of the Second World War Paweł Machcewicz:
If it is possible, I would simply like to answer the questions of the members of the Sejm.

Chairwoman Elżbieta Kruk (Law and Justice):
Yes, please.

Director of the Museum of the Second World War Paweł Machcewicz:
I will try to be systematic, starting with the most substantive issues. I want to reassure Mr. Czartoryski. All the topics you mentioned—Monte Cassino, Żegota, the 303 Squadron, the Polish Underground State, Jan Karski,[13] Katyń—are on display, and some of those things that you accuse us of not including are even in these dozen or so pages of material [that we provided]. Please note that the uniform of a 303 Squadron pilot is shown here. As for the rescue of Jews, there is an exceptional, unique object, a bowl belonging to a Polish family hiding Jews. We also tell you about Żegota, but we want to talk about ordinary people who ... If you had read this carefully, there is a description... . Please do not reproach me for not saying the truth, because it is simply offensive. You have made a very selective review of this material. For example, you did not write about the fact that in the material you read there is information about Katyń and that we are the first museum in Poland and the world that shows crimes against the Poles in Volhynia at the hands of Ukrainians. We did this in a very moving way, by showing objects exhumed from mass graves. You are accusing us of not including concentration camps. There are objects from Mauthausen-Gusen, which was the place of martyrdom of the Polish intelligentsia. Are we to talk about all concentration camps in such a short space? Nota bene: Auschwitz is also there.

A very large part of our narrative is the Polish Underground State. There will not be any other museum in Poland (even the Museum of Polish History, because I think there will not

---

13 An envoy of the Polish Underground State who shared with Western leaders in 1943 information about the extermination of Jews in occupied Poland.

be enough space) with such an extensive story of the Polish Underground State as there will be in our museum. A few hundred square meters about the Polish Underground State itself. Ladies and gentlemen, I would like to assure you that from the very beginning, the goal of creating this museum was to present the Polish historical narrative to the world by inscribing Polish sensitivity, Polish experience into a world story. For the first time, this museum will tell in such a broad way that Poles have become victims of two aggressions, not only German but also Soviet. Mr. Czartoryski somehow did not want to mention that this document, which he read, includes objects showing Soviet aggression—for example, the Polish border sign, the Polish eagle dropped into the Zbrucz River. You have the right to criticize, but I would ask that this criticism be honest, because any project can be slandered and criticized by being selective. I'm not saying that I'm right in everything, but I'm asking for honest criticism, not lynching.

There is no museum, no museum currently in existence that would show Polish fate and Polish historical experience to such a large extent as do our exhibitions. Of course, there is no unanimity. I agree with Professor Żaryn that there is an ideological dispute, but please do not deny that we present the Polish vision of history. It may differ from your vision, but you do not have a monopoly on Polishness; you do not have a monopoly on patriotism. Someone spoke about cosmopolitism. I want to say that I am a Polish historian who has devoted his entire professional life to dealing with history, and I think that such an accusation is out of place. Ladies and gentlemen, when it comes to ...

Chairwoman Elżbieta Kruk (Law and Justice):
Professor, please, only the most important things.

Director of the Museum of the Second World War Paweł Machcewicz:
Madam President . ..

Chairwoman Elżbieta Kruk (Law and Justice):
I'm sorry, the next committee is waiting, in which participate . ..

Director of the Museum of the Second World War Paweł Machcewicz:
You let Senator Żaryn speak for fifteen minutes, let me . ..

Chairwoman Elżbieta Kruk (Law and Justice):
Professor, you spoke very long at the beginning. You, Sir, had time to say what you have to say... .

Director of the Museum of the Second World War Paweł Machcewicz:
We show the suffering of the civilian population, because the main part of the Polish victims of the Second World War was the civilian population. But of course we also show Monte Cassino and the route of the armored division of General Maczek. The tank, which Ms. Sobecka asked about, is painted in the colors of General Maczek's armored division . ..

MP Anna Sobecka (Law and Justice):
Covered with a concrete slab.

Director of the Museum of the Second World War Paweł Machcewicz:
MP, this tank was a target for the Belgian army on a training ground; it was in pieces, shot. We imported it on a flatbed to Poland; we renovated it. We put in the engine of the Jelcz bus —a bus, because we did not have a Sherman engine—and it was promoting the museum for

several years. Now is the time to create the exhibition. The tank cannot both drive and be part of the exhibition. At this exhibition, it talks of the combat route of General Maczek's armored division, and it is probably a worthy story to tell. It is completely safe.[14]

I have devoted so much space to the meeting of the Sejm Culture Committee because it best reflects the mentality and arguments of the people who had decided not to allow the Museum of the Second World War to open in the shape in which we created it. Personally, I found it striking, this conviction of politicians from the ruling party that they had the right to decide the detailed nature of the exhibition, even which artifacts could be displayed in it. I also found striking their belief that they had a kind of monopoly on Polishness, that they could decide what was in line with Polish national sensitivity and what was not. On a deeper level, their arguments reflected the fundamental conviction that as a nation we are in jeopardy and under constant attack, our historical identity and merit undermined. We should only show our own history, and any attempt to incorporate it into the wider context and take into account the experiences of other nations threatens Polish uniqueness and perhaps Polishness itself. The accusations of "universalism" and "cosmopolitanism" were no accident.

After seeing the record of the committee's hearing, historian Piotr Osęka, who researches the nationalistic and anti-Semitic propaganda of 1968, pointed out to me that it reminded him, like the earlier attacks on the Museum, of the way in which the so-called encyclopedists were brought down by the Communists. At that time, a multimonth press campaign aimed at the editors of the great encyclopedia published by the Państwowe Wydawnictwo Naukowe (National Scientific Publishing House) ended with the dispersal of the greater part of the editorial office.[15] The starting point was the critique of entries devoted to concentration and extermination camps, the content of which supposedly diminished the martyrdom of the Polish nation and overly stressed the extermination of Jews. Then, almost all the entries in the encyclopedia about the Second World War were crushed under heavy criticism.

Here is only one example of the accusations from that time. A journalist from the magazine *Prawo i Życie*, Tadeusz Kur, wrote in his article titled "Encyclopedists," "Unfortunately, it is necessary to accuse the editors of the [Great Encyclopedia] that the struggle and martyrdom of the Polish nation was not an issue that was in their heart... . It is high time; put an end to this little adventure; make

---

14 Kancelaria Sejmu. Biuro Komisji Sejmowych. Sejm Rzeczypospolitej Polskiej. Pełny zapis przebiegu posiedzenia Komisji Kultury i Środków Przekazu (24), June 8, 2016, www.sejm.gov.pl.
15 See Tadeusz Rutkowski, *Adam Bromberg i "Encyklopedyści". Kartka z dziejów inteligencji w PRL*. (Warszawa: Wydawnictwa Uniwersytetu Warszawskiego 2010).

decisions that prevent similar phenomena for the future; apply measures to repair the damage so far." Tadeusz Kur firmly rejected dividing Nazi camps into extermination and concentration camps and providing information that the fate of Jews was more horrible than that of Poles. "All Nazi camps were extermination camps. The Polish experience speaks so clearly to it." He mentioned dozens of places of execution of the Polish population that were not included in the Great Encyclopedia, as well as numerous places of partisan warfare. In Kur's view, the pages of the encyclopedia "reduced the role of the country that fought" and "underestimated the underground armed struggle of Poland against the Nazi occupier." In turn, "Germans were the lucky ones in the Encyclopedia and they had the grace in the eyes of its editors." This was a result, among other causes, of emphasizing the main role of the SS in the Nazi terror machinery. "Thus, the government of the Third Reich, the Wehrmacht, and the civil administration in the occupied countries received a rehabilitation certificate from the editors of [the encyclopedia]."[16]

And one more example, a fragment of an interview given to the Polish Press Agency by Mieczysław Moczar, an influential minister of internal affairs and, at the time, the leader of the nationalist faction of the Communist authorities:

> Is the state of knowledge about the attitude of Poles in the years of the past war sufficient in our society?
>
> It is absolutely insufficient. The war matters were, so far, one-sidedly, often dishonestly presented. Omitted on purpose were some matters important from the point of view of tradition and national dignity. The best example of this is the Great Encyclopedia, developed under the aegis of people known especially from the Stalinist period. They did not care much about matters that constitute the pride of the Polish nation... . [T]hey tried to distort our history and weaken the memory of national traditions. Everything that our nation accomplished during the last war requires a dignified commemoration, passing it to the young generation. It cannot be depreciated by shallow pseudo-analysis. These matters must be dealt with by people who understand this, who feel the breath of our nation.[17]

One could replace Stalinists with "people who serve the European elite, Brussels and Berlin," and everything else remains valid. What is missing is the contemporary term "pedagogy of shame," which Moczar could not have known. After several dozen editors of the Great Encyclopedia were fired, an errata was prepared

---

16 Tadeusz Kur, "Encyklopedyści," *Prawo i Życie*, March 24, 1968, quoted in Piotr Osęka, *Syjoniści, inspiratorzy, wichrzyciele. Obraz wroga w propagandzie marca 1968* (Warszawa: Żydowski Instytut Historyczny, 1999).

17 *Należy przekazać młodzieży historię naszych heroicznych zmagań. Wywiad z Min. Mieczysławem Moczarem*, April 12, 1968, quoted in Osęka, *Syjoniści, inspiratorzy, wichrzyciele*, 233.

and inserted into the next volume of the encyclopedia, which appeared in December 1968. It contained a new version of the entry on "Nazi camps" (no longer divided into extermination camps and concentration camps). Sixty lines of text about the martyrdom of Poles in concentration camps were added, and information about the extermination of Jews was reduced by half, to fifteen lines. It was also decided that the second edition of the encyclopedia would significantly increase the number of entries devoted to Polish subjects.[18]

---

18 Tomasz Stańczyk, "Atak na Encyklopedystów," *Rzeczpospolita*, October 13, 2008; Rutkowski, *Adam Bromberg i "Encyklopedyści,"* 264.

Reviewers Unmasked

After meeting with Minister Piotr Gliński, I had repeatedly asked him publicly to disclose the reviews of our exhibitions that he commissioned. We really did invite the debate. It was worthwhile; we were open to criticism—I was speaking in June 2016, during the session of the Sejm Committee on Culture.

The appeals were ineffective, and the whole situation became increasingly reminiscent of the atmosphere of Franz Kafka's *The Trial*. Our exhibitions had been subjected to ruthless criticism for months, mostly from people who did not have any idea about them. There were four mythical reviews, but it was not known by whom they had been written or specifically what the criticism leveled against us was. One could only guess that the accusations were very serious, especially as the minister of culture formulated his opinion of the exhibitions based on the reviews. The parliamentary debate indicated their possible nature.

Once it became clear that Deputy Prime Minister Gliński did not intend to comply with the promise made to me and would not voluntarily reveal the reviews, the mayor of Gdańsk, Paweł Adamowicz, requested them under the access-to-information law. At the beginning of July 2016, the Gdańsk authorities received three "declassified" reviews (the fourth "dissolved" somehow; it is not known whether it existed at all) and gave them to both the Museum and the media. The situation at that point was peculiar. The Ministry of Culture apparently did not warn the reviewers that they would have to face public opinion. Piotr Niwiński, asked by a journalist from *Gazeta Wyborcza*, denied that he had written a review at all: "And where did this information that I wrote such a review come from? You, Sir, surprise me and flatter me at the same time."[19]

I studied the reviews thoroughly and immediately placed them on the Museum's website. We were most amazed that the reviewers used only one short document from among the extremely extensive package that we had submitted to the Ministry of Culture in early January. The package included so-called intervention tables, on the basis of which it was possible to reconstruct every part of the exhibition, and a film about the exhibition (narrated by Maja Ostaszewska) showing numerous artifacts that visitors would see. One had to know how to use the intervention tables, but in a letter to Deputy Minister of Culture Jarosław Sellin, I declared that we would be happy to explain everything and answer any questions about the exhibitions. Nobody approached us, and the reviewers relied only on the seventy-five-page, richly illustrated "Outline of the Exhibitions."

---

[19] Krzysztof Katka, "Znamy recenzje wystawy Muzeum II Wojny Światowej. Źle postawione akcenty," *Gazeta Wyborcza Trójmiasto*, July 11, 2016.

When, on behalf of Sellin, one of the officials asked me for materials on the subject of the exhibitions, he clearly indicated that in addition to other materials, we needed to provide the ministers with something short, a summary, so that they would have time to read it. So we did, but it did not even occur to us that the verdict on the Museum would be issued only on the basis of the short "Outline of the Exhibitions" (which was also provided to the members of the Committee on Culture) without an examination of the details of the exhibitions. On the first page of the outline the reviewers could read that the exhibitions "cannot be fully presented without proper technical documentation. When analyzing this document, one should therefore use the plans for the whole exhibition and its individual parts, technical sections, and the so-called intervention tables in which all the components of the design were described."[20]

This statement was completely ignored. The reviewers charged that the exhibition lacked many important threads from the wartime fate of Poland; meanwhile, these threads were there, but the reviewers could not or did not want to find them. Some of the threads were included even in the "Outline of the Exhibitions" on which they based the reviews, but apparently they did not carefully study even that. Therefore, the major part of each review consisted of a long litany of completely misguided allegations, listing the alleged deficiencies and omissions of the exhibition.

Jan Żaryn accused us of omitting, among others things, the bombings of Polish cities by the Germans in 1939, the defense of Warsaw, and the figure of Julien Bryan, an American correspondent who spent several weeks in the besieged Warsaw and whose pictures and films were later watched by the whole world. In fact, not only were these topics included but each of them had a dedicated room. In the case of Bryan, it was even a cinema room in which visitors watched fragments of his film *The Siege*. Also, they could admire an authentic film camera that belonged to the director, which, after several years of persuasion, was handed over to me by Sam Bryan, the son of the American documentary maker. I personally brought it to Gdańsk in a suitcase from New York.

Piotr Semka's allegations were perhaps the most bizarre because he accused us of omitting topics that were mentioned in the documents that we provided and on which he based his review. Semka claimed that in the exhibitions "there is not a word about the massacre in Wola" during the Warsaw Uprising and also that we ignored the crimes committed against the Poles in Volhynia.

---

[20] Paweł Machcewicz et al., *Muzeum II Wojny Światowej w Gdańsku. Program funkcjonalno-użytkowy wystawy głównej*, Gdansk.pl, January 2016, https://www.gdansk.pl/download/2016-04/71915.pdf.

In fact, the exhibitions did show the slaughter in Wola and the extermination of Poles in Volhynia—the latter in a dedicated room. He did not notice that both topics were noted in the "Outline of the Exhibitions" in which we even included photos and descriptions of relevant artifacts: "a child's shoe belonging to one of the victims of the slaughter of civilian residents of Warsaw during the Warsaw Uprising" and "medallions found in the graves of Poles murdered by the Ukraine Insurgent Army in Ostrówki and Wola Ostrowiecka."

"Somehow, however, the Battle of Monte Cassino disappears," Semka raged. "Yes! This is not a mistake—in the design of the exhibition of the Polish Museum of the Second World War there is no, unless I missed something, not a word about Monte Cassino." Indeed, he did miss something. Monte Cassino and all other battles in which Polish soldiers participated were shown in the exhibition, and the II Corps, fighting on the Italian front, was mentioned in the said "Outline of the Exhibitions," in which we posted pictures of personal objects of General Władysław Anders, its commander, illustrating this story.

In turn, Piotr Niwiński claimed that the exhibition was not modern. "The interactivity is practically nonexistent (I counted three stations, altogether)." A little further on, he wrote, "I appeal for interactivity once more." It was very important to him, and he repeated this issue many times in what was not, after all, a long review. In fact, there were 240 interactive multimedia stations on display, and each of them was described in the materials we provided to the Ministry of Culture.

For us, after working so many years on the exhibitions, something else was striking: the reviewers' complete misunderstanding of what a museum is, how its narrative is built, and the role of the artifacts. None of the three reviewers had any museum experience, and this was evident in almost every sentence they wrote. They treated the exhibition as a large book or even an encyclopedia; they demanded the inclusion of a huge amount of information, apparently not understanding that a museum story is built through examples, symbols, and objects.

More important were the allegations of a general nature, which reflected a real ideological dispute and a completely different understanding of both history and the role of museums. One of the most important was the allegation that we devoted too little attention to the fate of Poland, that Poland was apparently "obscured" by other nations. It was in fact the same thread that had figured already in the attacks on the Museum in 2008 (Semka and Żaryn also played a leading role then) and during the above-mentioned meeting of the Sejm's Committee on Culture. Jan Żaryn concluded in his review, "The authors made a strategic choice, in my opinion, in which the Polish point of view on history was 'buried' in pseudo-universalism, a version of universal history." These were essentially the same charges he had hurled during the Sejm's committee hearing: a lack of sensitivity to and detachment from national tradition. "The authors, therefore, tried to send a signal that identification with the drama of the Second World War was possible if we approached this period on an individual level, without referring to national myths, the memory of the collective experience of Poles. Of course, no one proposes such an operation when analyzing the Holocaust, to which only Jews were subject. As if the Poles did not have their own experience (including memories and myths) and it was not worth showing this on Polish soil." The review ends with his own concept of what the museum should first of all show: "who we are today and why we are the way we are: freedom lovers, Catholics, patriots, etc., and above all proud of our history."

Piotr Semka also accused us of harmful "universalism" and sent us to Brussels, just as Jan Pospieszalski had done in his parliamentary speech:

> One could understand this concept if it was a museum in Brussels about the fate of Europe. But this museum is to be created in a country that is often considered to be Hitler's helper in the Holocaust. Such a country should care about commemorating the glory of Polish arms and the martyrdom of Poles. Only when we make sure that a tourist from London or Vienna gets to know our contribution to the defeat of Nazism, will it be possible to expand the exhibition with more universal themes. It is also difficult to remain silent about the fact that presenting the Second World War as the anonymous suffering of all Europeans is beneficial to the Germans and the nations that collaborated with the Third Reich.

In the exhibition, Semka saw "intrusive reminders of Germans as victims at every step." He unequivocally defined the place of the Museum's creators on the ideological map of Poland: "The distrust of some of our elites toward the concept of historical policy proposed by the Warsaw Rising Museum seems almost absurd. The degree of suspicion about this concept is amazing." He ended his review with strong conclusions: "But let the authors of the concept not be surprised that there is an expectation that the fate of our lands and our ancestors will be clearly brought to the forefront... . The present concept meets the expectations of the Polish Second World War story only to a limited extent. The exhibition requires a significant remake."

I have always been impressed by the self-confidence of people who grant themselves the right to decide what is the "Polish point of view" (like Jarosław Kaczyński when he spoke about our Museum) and the "Polish story" and what is not. Others cannot present a different interpretation of Polish history or experience "Polishness" differently; they are simply excluded.

The reviewers also alleged that the exhibition was antiwar. As Niwiński wrote, "With the main emphasis placed on the horror of war, depicted through the fate of the civilian population, we get a rather one-sided picture. It has shown in a rather selective manner ... the dark side of the war, but what arose out of it is far away in the background. You can specify this as the dominance of unhappiness over other traits. It would be most appropriate to summarize the whole message with the phrase from the Communist era: No more war."

Both Niwiński and Semka demanded more military history. They reacted negatively to devoting so much attention to the sacrifice of the civilian population, by which the achievements of Polish soldiers were diminished, as was the notion that war "strengthens character."[21]

Rafał Wnuk and I prepared a response to the reviews, published in an abbreviated version in a daily newspaper, *Rzeczpospolita*. We pointed out that the vast majority of the allegedly omitted topics were in fact included in the exhibitions, but we also referred to more general allegations and issues. We explained why, as one reviewer charged, ours was a "museum of martyrdom" in which so much space was devoted to the suffering of the civilian population:

> During the war over 200,000 Polish citizens died fighting with weapons in hand; we recall their fate many times in the museum and we pay homage to them. Civilian losses were, however, incomparably greater. Over 5 million people were murdered: about 3 million Polish Jews and over 2 million ethnic Poles. They generally died a more terrible death than a soldier's death: in gas chambers, concentration camps, executions as in Pomerania in 1939, in Palmiry, or in Wola during the Warsaw Uprising. It astounds us that the creators of the museum being built in Poland, the country that suffered such terrible losses in the war, must explain themselves for adopting this perspective in the creation of the exhibitions.

We also reminded readers that the message "No more war" was not formulated by Communist propagandists but in fact had a deeply Christian meaning. Pope Paul VI proclaimed it during his first visit to the United Nations in 1965, and then John Paul II repeated the message. We remembered the words on the war spoken by the Polish pope in 2003: "The size of the losses suffered, and even more so, the amount of suffering inflicted on people, families and environments, is really difficult to calculate... . The war was not only on the front lines, but as a total war it struck whole societies. Whole communities were deported. Thousands of peo-

---

21 Piotr Niwiński, Kierownik Zakładu Nauki o Cywilizacji, *Recenzja programu funkcjonalno-użytkowego wystawy głównej przygotowanej przez Muzeum II Wojny Światowej w Gdańsku autorstwa Prof. Dr. Hab. Pawła Machcewicza, Dr. Hab. Piotra M. Majewskiego, Dr. Janusza Marszalca, Dr. Hab. Prof. KUL Rafała Wnuka, Stan na styczeń 2016*; Piotr Semka, *Recenzja dokumentu "Program funkcjonalno-użytkowy wystawy głównej" przygotowanej przez Muzeum II Wojny Światowej w Gdańsku przez zespół: Prof. Dr Hab. Paweł Machcewicz, Dr Hab. Piotr M. Majewski, Dr Janusz Marszalec, Dr Hab. Prof. KUL Rafał Wnuk. Dokument datowany na styczeń 2016*; Jan Żaryn, Historyk, *Recenzja wewnętrzna: Program funkcjonalno-użytkowy wystawy głównej. Muzeum II Wojny Światowej w Gdańsku*, 2016.

ple were victims of prisons, torture and executions. People who were beyond the theater of war died as victims of bombing and systematic terror, the organized means of which were concentration labor camps, which turned into death camps." Wnuk and I declared, "This vision of war, which Piotr Niwiński would certainly consider as too 'negative,' and the fate of the civilian population is close to the hearts of the creators of the exhibitions of the Museum of the Second World War."

We also pointed out that Piotr Semka and Jan Żaryn had fought the Museum of the Second World War from the very beginning, so the minister of culture had known in advance what the content of the commissioned reviews would be. These reviews had nothing to do with scholarly reliability, as Piotr Gliński, being a professor of sociology, had to know well; rather, they were an attempt to justify the planned steps against the Museum. At the end of our response we appealed to the minister of culture, though without any internal conviction, to allow us to open the museum as planned and to let public opinion be the judge.

> We live in a country ruled indivisibly by one political option, which obviously has a mandate coming from democratic elections. It can create all the facilities it promises, which will shape the awareness of Poles: the Museum of Polish History, museums of cursed soldiers (there are already two established), western lands,[22] and eastern borderlands.[23] Can there not be space in the museum landscape for the Museum of the Second World War in the form created by historians whose competence, regardless of the differences of views, no one has questioned?

> If the current government will not allow the opening of the exhibitions, already practically complete, this will be the most intrusive interference, unprecedented in a free Poland, in the autonomy of cultural institutions and the sphere of public history. This type of preventive censorship has not taken place since 1989. Sooner or later it will ricochet onto other institutions, including those created by the current authorities. After all, nobody [from the previous Civic Platform government] changed the exhibitions of the Warsaw Rising Museum, the oeuvre of Lech Kaczyński, whose director was an MP and politician of the Law and Justice Party. We are also inquiring if Minister Gliński really considers forced administrative solutions, characteristic of authoritarian systems rather than a pluralistic democracy, as the best way to build Poles' historical consciousness.[24]

A statement by Minister Gliński's in Sejm can be considered a response: "The content of the exhibition of such an important museum as the Museum of the

---

22 The territories that Poland took over from Germany in 1945.
23 The territories that Poland lost to the Soviet Union in 1944 and 1945.
24 Paweł Machcewicz and Rafał Wnuk, "Muzeum zwykłych ludzi," *Rzeczpospolita*, August 7, 2016.

Second World War is not a matter of the tastes of even outstanding Polish and foreign historians or the tastes of ministers. The content of this exhibition should express the basic assumptions of historical policy in accordance with the Polish raison d'état... . The concept must articulate, in a deeper sense of this word, the position of these politicians who currently have a social mandate regarding historical politics."[25]

By this logic, those who win elections have the right to change the content of museum exhibitions and maybe also school textbooks. Such an understanding of the public role of history and museums deprives them of any autonomy from the world of politics. This autonomy, after all, is one of the important elements of the democratic state in which we have lived since 1989, distinguishing it from the times of the Communist dictatorship. Responding to arguments such as those presented in the Sejm by Minister Gliński, I repeatedly asked the rhetorical question, Do we want to live in a country where after each election, a new government, having a social mandate, changes the content of exhibitions and even removes uncomfortable museums? I also proposed an intellectual exercise for supporters of the merger of the museums. I asked them to imagine a situation in which the mayor of Warsaw, a politician of the Civic Platform, creates a "Museum of the 1944 Rising in Wola," which undertakes no activities. After a few months, someone argues that it is not rational for two museums on the same subject to operate in one city. The newly created museum is to be merged with the Warsaw Rising Museum, associated by the general public with the Law and Justice Party. The director of the latter loses his job, and the exhibitions, criticized many times, are to be changed. This scenario elicited no reaction from supporters of the forced merger facing the Museum of the Second World War, perhaps because they were convinced that Law and Justice would rule forever and did not believe that the mechanism, once put in motion, could also be used in the future against them.

I invited the reviewers and also Minister Gliński to Gdańsk for a public debate about the exhibitions, sending them the newly published catalog and urging them to familiarize themselves with the actual content of the exhibitions: I reminded them of detailed materials that we had already forwarded to the Ministry of Culture and which evidently were not read by the reviewers. By the beginning of September, they had declined. This did not surprise me much. It was easier to pass covert judgments than to publicly confront the arguments of his-

---

25 "Gliński: Wystawa Muzeum II Wojny Światowej musi być zgodna z polską racją stanu," Polska Agencja Prasowa, September 14, 2016, https://dzieje.pl/aktualnosci/glinski-wystawa-muzeum-ii-wojny-musi-byc-zgodna-z-polska-racja-stanu.

torians who had worked on the exhibitions for eight years, and to do that just before the opening of the Museum, which was now so uncertain. There was also no answer to my and Rafał Wnuk's article in *Rzeczpospolita*. The Ministry of Culture, however, stood by the reviewers and issued an official statement. It praised the research achievements of the reviewers; our answers were called "unreliable." It claimed, "Public statements that the reviewers did not use all the materials necessary for the review are untrue." The official statement was not without a warning: "The minister of culture and national heritage calls for the preservation of integrity in the debate about the Museum of the Second World War... . Offending recognized historians in public does not fall within the limits of such a debate."[26] This was yet another interesting contribution to what a "reliable" debate should look like and who should define its rules.

The revelation of the content of the reviews caused quite a stir in public opinion. Allegations that the Museum showed too much suffering, that it painted too "black" a picture of the war with too few "bright," "positive" sides, and that the war's "character-building" qualities were ignored, provoked numerous commentaries. In an article titled "Oh, How Pretty Is the War," renowned journalist Marcin Meller wrote,

> For almost a decade I have described various wars, from the Caucasus through ex-Yugoslavia to Africa... . I never had the right words to describe the hell of war, the fate of people lost in the slaughter, murder, and extermination. And maybe that is why I am completely uninterested in various historical and war reenactments that seem to me infinitely infantile and mindless. And I cannot stand resolute and optimistic tales about the Warsaw Uprising. (Which didn't we almost win! Only, bad luck, some 200,000 people happened to die.) Because there is nothing beautiful in war. Because war is not *The Four Troopers from a Tank Division*[27] or even *Time of Honor*[28] but a dehumanizing hell. And now I read that the Law and Justice reviewers employed by the minister of culture, Piotr Gliński, muddle the idea of the Museum of the Second World War in Gdańsk because "the museum carries mainly the message of the war as exceptional tragedy. No positive features are exhibited, such as pa-

---

[26] "Oświadczenie MKiDN w sprawie nierzetelnych zarzutów kierownictwa Muzeum II Wojny Światowej," MKiDN, July 18, 2016, http://www.mkidn.gov.pl/pages/posts/oswiadczenie-mkidn-w-sprawie-nierzetelnych-zarzutow-kierownictwa-muzeum-ii-wojny-swiatowej-6456.php;
"MKiDN: Recenzenci wystawy Muzeum II Wojny—rzetelni i rzeczowi," Polska Agencja Prasowa, July 19, 2016, https://dzieje.pl/aktualnosci/mkidn-recenzenci-wystawy-muzeum-ii-wojny-rzetelni-i-rzeczowi.

[27] A Polish series about the wartime adventures of four young men and their dog, Szarik, serving in the 1st Polish Army in 1944 and 1945, which fought alongside the Red Army. The series, produced between 1966 and 1970, attained enormous popularity and was frequently shown in reruns until 1989.

[28] The series *Time of Honor*, shown between 2008 and 2013, told the story of Polish paratroopers who were trained in England and fought as part of Warsaw's underground resistance.

triotism and sacrifice." How many times have I heard from ideologues and those inflicting slaughter in various corners of the world about these "positive traits" of war, the "hardening of man," about which Niwiński, Semka, and Żaryn care so much.[29]

Those of the generation that had survived the war also spoke. Joanna Muszkowska-Penson said,

> This is a harmful sentence. I saw the entire Second World War in its worst acts... . As early as March 1941 I was arrested in my hometown of Warsaw for conspiracy in the Union of Armed Struggle. I went through interrogations and beating at Pawiak prison and Gestapo headquarters at Szucha Street. I came to the Ravensbrück camp, in one of the first transports, where I remained for four years until its liquidation. Most of my friends did not survive. War destroys everything. It is total destruction, including psychological. Why is one young man supposed to kill another whom he does not know and who did nothing to him? War is always deeply immoral and unjust. The best die, those most willing to sacrifice, those in the front line. There is nothing as bad as war.[30]

Aleksander Tarnawski, aka "Upłaz," the last Polish *Cichociemny*[31] still alive, was blunter than Professor Joanna Penson. He encouraged Piotr Semka and Piotr Niwiński to go to Syria, join the troops of one of the fighting factions, and get to know what war was and how it "strengthens."[32]

Certainly not everyone agreed with such assessments, and the reviewers were not alone in their notion of what war was and how it should be shown. Over the preceding decade, renowned poet and writer Jarosław Marek Rymkiewicz had become the bard of the Right. In his book on the Warsaw Uprising, he put forward the thesis that this massacre had cemented the Poles as a nation, allowing them to survive the postwar decades. Throughout Rymkiewicz's book one can see the fascination with death, suffering, and sacrifice of life for the homeland.[33]

Human losses and suffering were by no means central to the story of the Warsaw Rising Museum, visited annually by several hundred thousand people. Regardless of the message of its exhibitions, from the very beginning the museum conducted many promotional and educational activities that created the

---

29 Marcin Meller, "Jak to na wojence ładnie," *Newsweek*, July 18–24, 2016.
30 "Nie macie prawa niszczyć Lecha. Z Joanną Muszkowską-Penson Rozmawia Marek Górlikowski," *Gazeta Wyborcza*, February 4–5, 2017.
31 The so-called *Cichociemni*, or "Silent Unseen," were elite paratroopers of the Polish Army in Exile, trained in Great Britain during World War II to conduct special operations in Nazi-occupied Poland.
32 *Wojna o Muzeum Wojny*, TVN 24, July 29, 2017.
33 See Jarosław Marek Rymkiewicz, *Kinderszenen* (Warszawa: Wydawnictwo Sic!, 2008).

image of the uprising as a joyous experience of young, beautiful people who should find imitators among contemporary youth. "For 11 years the museum has shown the uprising as a great adventure, and it will continue to show it as such over the next years," wrote a journalist in the renowned Catholic weekly *Tygodnik Powszechny* to mark an anniversary of the outbreak of the uprising.[34] One should add to this pop culture and even the commercialization of the memory of the uprising: a flood of coloring books, puzzles, mugs, T-shirts, and other ephemera with symbols of the upraising. Some of them are sold in the Warsaw Rising Museum. This certainly is not Tomasz Łubieński's perspective in his book on the uprising: "Not belonging in this legend: the girls raped in Zieleniak concentration camp, the boys shot on Narutowicz Square, the wounded murdered in hospitals, the children, women, and elderly killed just because they could not hide from the bombs or, if they did manage to get to the basement, were buried alive there or torn apart by grenades that the enemy threw through basement windows."[35]

A separate phenomenon, one on a European scale, is the growing popularity of the historical reenactment movement in Poland. Thousands of people put on uniforms, take up dummy arms, and act out battle scenes from the war, often the Warsaw Uprising, or episodes from the history of the "cursed soldiers." Thousands of others watch those television shows that generally depict war and struggle as something very attractive and spectacular, revealing the best human characteristics. In this very superficial image of the war, there is no room for reflection on the price that is paid for resistance and heroism; wounds and death are purely orchestrated and theatrical. There is room, however, for fascination with militaries, uniforms, and weapons.[36] Even the Catholic Church in Poland felt disturbed by the scale and the nature of this phenomenon. The Council for Social Affairs at the Polish Episcopate published a document on the Christian shape of patriotism, which among other things addressed the communities of historical reenactors: "When preparing such productions, one should remember the *mysteria* of human death and suffering, fear and heroism, whose dignity and mystery cannot always be properly portrayed in mass, open-air presentations... . The staging, necessarily symbolically simplified, cannot express the whole

---

34 Kalina Błażejowska, "Powstańcze przebieranki," *Tygodnik Powszechny*, August 14, 2016. See also the conversation with Jan Ołdakowski in "Zakładanie opasek to nie hołd," *Gazeta Wyborcza. Magazyn Stołeczny*, July 29, 2016.
35 Tomasz Łubieński, *Ani triumf ani zgon* (Warszawa: Wydawnictwo Ministerstwa Obrony Narodowej, 2004), 42–43.
36 See Tomasz Szlendak, ed., *Dziedzictwo w akcji. Rekonstrukcja historyczna jako sposób uczestnictwa w kulturze* (Warszawa: Narodowe Centrum Kultury, 2012).

drama, and sometimes the cruelty of historical situations." Another suggestion by the episcopate was much more universal and somewhat referred to the dispute about the depiction of the war in our Museum: "War, although it often reveals human greatness and heroism, is not a colorful tale or adventure, but a drama of suffering and evil that should always be prevented."[37]

On top of all this, there was the cult of the "cursed soldiers," the main element of the historical policy of Law and Justice from the moment it took power in 2015, but which was also evident even earlier among the Right.[38] In every possible way, politicians of the ruling camp and the media they controlled promoted the soldiers of the postwar underground as a model for today's young Poles, excluding all other, especially "civilian," ways of serving the homeland. Countless concerts, festivals, and school assemblies were devoted to the "cursed." Films were made about them; they become patrons of Territorial Defense units, voluntary training groups that support the regular army. All that seemed to remain of the complicated reality of the 1940s was the imperative that after the end of the war, one should have gone to the forest with weapons to fight against the invader.

This understanding and emotional perception of history must have clashed with the message of our Museum, with its nonmilitary perspective focused on the experience of the civilian population, showing the war above all as a terrible tragedy. Of course, we showed it also as a time of the highest heroism and dedication, which the reviewers and other critics did not want to or could not see. In essence, this dispute had much deeper roots and was embedded more in Polish tradition than in the ideological conflicts of the last dozen or so years. Maria Janion, an eminent Polish cultural historian, wrote about a romantic cult of struggles and uprisings that dominated Polish consciousness in the nineteenth century, characterizing it as "the conviction of the superiority of that which is 'military,' 'heroic,' and 'poetic over that which is 'civilian,' 'ordinary' and 'mundane.'" As Janion pointed out, "The deeply anticivilian and romantic orientation of public awareness worried those who during the partition of Poland in the nineteenth century chose antimilitary, cultural, and economic endeavor."[39] However, this romantic cult of an armed struggle occupied a central place in the Pol-

---

[37] "Chrześcijański kształt patriotyzmu. Dokument Konferencji Episkopatu przygotowany przez Radę ds. Społecznych," Konferencja Episkopatu Polski, April 27, 2017, https://episkopat.pl/en/chrzescijanski-ksztalt-patriotyzmu-dokument-konferencji-episkopatu-polski-przygotowany-przez-rade-ds-spolecznych.
[38] See Rafał Wnuk, "Wokół mitu 'żołnierzy wyklętych,'" *Przegląd Polityczny* 135 (2016): 184–187.
[39] Maria Janion, *Płacz generała. Eseje o wojnie* (Warszawa: Wydawnictwo Sic!, 1998), 25–27.

ish historical memory and understanding of patriotism, which the Second World War only strengthened.

Attempts to broaden this memory and this understanding to include the experiences of the civilian population have usually met with resistance, shock, and disappointment. Stefan Chwin has even written about the taboo of the "civil voice" in Polish culture and literature.[40] A telling example was the attacks on a poet, Miron Białoszewski, after the publication of his *Diary of the Warsaw Uprising* in 1970. He was reprimanded for not fighting (in fact he volunteered but was not accepted into the Home Army), for trying to survive. He was ridiculed as an "outsider" and called "Mironek" (a diminutive of his first name), a little child who does not understand anything about the fight surrounding him and is only preoccupied with his own mundane matters. The testimony given by the author was irrelevant to his critics, Maria Janion noted. "Białoszewski lived through the entire 63-day long Gehenna of the Warsaw residents. He was not spared much (for example, he was not used as a human shield by German tanks attacking the positions of insurgents, but he did hear a lot about that from those who were). It is impossible to enumerate all the martyrdoms of the population—the *Diary* mostly consists of them. In this doomsday world, a sobbing child is calmed down with the words, 'Do not cry, you will not live anyway.'"[41]

Still, all this was too mundane, less valuable and patriotic than the armed struggle, in which the narrator and other heroes of the *Diary* did not participate, although they were in the middle of it, their lives threatened by it. A similar attitude could be found in those criticizing and even discounting our Museum as a "museum of martyrdom" or "Museum of the Fate of the Civilian Population During the Second World War," as one reviewer wrote. Białoszewski's perspective was considered lesser and inferior, and Rymkiewicz's vision appealed more to the imagination. In his book *Kinderszenen*, the latter warned Poles that they should be ready to fight, when the German Tiger tanks again, as in August 1944, appear in the Warsaw neighborhood of Krakowskie Przedmieście.[42]

And what will happen if the Tiger tanks do not show up? Will we just wait for them, and will this readiness be the main measure of patriotism? In our Museum, one of the most extensive sections was the story of the Polish Underground State, a grassroots movement, a result of thousands of social initiatives conducted mainly by civilians. This phenomenon of self-organization—service for the common good, not necessarily with a weapon in hand—is still insuffi-

---

40 Stefan Chwin, "Strefy chronione," *Res Publica*, October 1995.
41 Janion, *Płacz generała*, 127.
42 Rymkiewicz, *Kinderszenen*, 65.

ciently present in Polish historical memory, obscured by the story of military struggle. Nor is it a strong component of the contemporary dominant model of patriotism, in which mundane work for the community is not as highly valued as the hypothetical readiness to die for it.

Stalingrad Instead of Blitzkrieg

The modified announcement of the minister of culture about the "merger" of museums (under the name of the Museum of the Second World War) was released on May 6, 2016. On August 5, after the three-month waiting period, our Museum could be liquidated. As the date approached, the tension grew. We did not have any information from the Ministry of Culture as to what exactly would happen. As a rule, Minister Gliński and Deputy Minister Sellin did not reply to any of my letters, and Gliński did not address the question that I asked at the June meeting of the Sejm: What would the future of the Museum be and what would happen to its employees? Before August 5, I packed all my things to be ready to leave the Museum when the guillotine dropped. On doomsday, I was sitting in an empty office, waiting for news from the Ministry of Culture and receiving dozens of phone calls from journalists who were ready to report on the last moments of the Museum of the Second World War. I could feel the Museum employees carefully looking at me when they came in with some issue or just passed by the office. They were waiting for any information; they were probably also curious about how I could withstand the growing pressure and how I would behave at the critical moment. However, nothing happened on August 5 or on subsequent days. There was also no information from Warsaw. Journalists began to storm the ministry with inquiries, which in response issued an enigmatic message that regulations "do not impose an obligation to issue an act on the merger of these museum facilities immediately after the lapse of three months from the announcement of the merger plan... . After three months, the Ministry of Culture and National Heritage may take steps to merge museums, including determining the date of the merger. Currently, the minister of culture and national heritage is preparing to issue the relevant act."

And that was all we knew about our future. We were to continue to work, make decisions about spending tens of millions of zlotys, take on responsibilities, and expect that at an unspecified moment, probably soon, all this would be terminated. In August, installation of the exhibitions, which was one of the most important phases of the entire undertaking, was just beginning in the museum building. I do not think that any other large public investment, in such a crucial moment of its creation, has ever been treated similarly, not only in Poland but anywhere in the European Union, by the government of its own country.

In any event, it seemed that we had no more than a few weeks ahead of us, perhaps due to the summer holidays, which even those at the top take. From our point of view, it was important that during these few months after the announcement of the liquidation of the Museum, the catalog of the main exhibition be published and distributed, the reviews disclosed, and the public discussion of the shape of the Museum be completed. The key was to begin installation of the exhibitions, which made it much more difficult and expensive to change them, if the government of Law and Justice really decided to do that.

In this very stressful moment of total uncertainty, support came from a completely unexpected source. Andrzej Nowak asked Timothy Snyder to issue a joint appeal to the authorities to allow us to open the Museum with the exhibitions that we had prepared. The appeal was published on August 13, appearing simultaneously in many newspapers and on internet portals, both those critical of the government and those supporting it, including the media that had joined the campaign against the Museum. In their appeal, Nowak and Snyder wrote,

> It is natural that historians differ in the interpretations of the past as well as in the assessments of today, which also applies to both of the undersigned. In the same way, historical museums in Poland and other countries present various visions of the past. Neither of us would design the Museum of the Second World War exhibition exactly as it is today. However, we do recognize that its exhibitions reflect both the historical truth in the dimension of the general picture of the war and the special place in it of Poland. We are in agreement that the Museum of the Second World War, in its present form, would create a unique opportunity for Poles to learn about the war outside Poland and for foreign visitors to learn about Polish history. As a member of the Academic Advisory Committee of the Museum of the Second World War (Snyder) and as a Polish historian unrelated to this undertaking (Nowak), we want to express our common hope that the Museum and its exhibitions will be completed and opened in accordance with existing plans.[43]

Snyder, a professor at the Yale University, is one of the world's best-known and most respected historians of the Second World War, totalitarianism, and Central and Eastern Europe. Nowak, a professor at the Jagiellonian University and the Institute of History of the Polish Academy of Sciences, is an expert on Polish-Russian relations and the author of valued works concerning mainly the nineteenth century and the first two decades of the twentieth century. Moreover, he is one of the main intellectual authorities of the political Right. He has written a lot about the desired model of patriotism and historical politics, supporting both its conservative form and often the Law and Justice Party directly. He has also been one of the harshest and most distinctive critics of the governments of the Civic Platform Party, accusing them of de facto abdication of Polishness and speaking very firmly against, among other things, their reform of history education in schools. His diagnoses were repeated by Law and Justice politicians,

---

43 "Prof. Andrzej Nowak i Timothy Snyder bronią Muzeum II Wojny Światowej: Wystawa ta oddaje prawdę historyczną," *Gazeta Wyborcza*, August 13, 2016.

and President Andrzej Duda appointed him to the National Development Council as the person responsible for the entire sphere of tradition, national memory, and historical policy.

The support of such a person in defense of the Museum of the Second World War was of great importance. His was the first dissent in the largely unified right wing in the battle over the Museum, and it came from a figure of great authority. This was a big surprise, and Nowak, despite all his achievements, did not escape the wave of hate on right-wing portals, where some accused him of betrayal. For me, his stand was also important from another point of view because of our previous dispute about the desired shape of historical memory, of which Westerplatte and Jedwabne were the symbols. This dispute was very often expressed in public as defining two significantly different and even opposing approaches to tradition, the public role of history, and the politics of memory.

Unfortunately, Gliński and Sellin completely ignored Nowak and Snyder's statement, as they had the earlier appeals of other historians or veterans. On September 6 the event that everyone expected finally happened. Minister Gliński signed the act on the "merger" of the museums and the creation of the "new" Museum of the Second World War. As before, we were not informed; the ministry announced the order on its website in the evening, and I learned about it again by phone from Janusz Marszalec, this time on a train traveling from Warsaw to Gdańsk. The "merger" was to take place on December 1, 2016, and Gliński specified the goals of the new museum he was establishing. These were to be primarily the

> building of collections and dissemination of knowledge on the history of the Polish Defense War of 1939 as the beginning of the global conflict—the Second World War—military and patriotic traditions and the history and achievements of the Polish Army of the Second Polish Republic, with particular emphasis on:
>
> a) the fate of the Military Transit Depot from the beginning of its creation to the end of the 20th century, including its heroic defense in September 1939;
>
> b) resistance of the Polish Soldier against the aggressors of the Third Reich and its allies and the Union of Soviet Socialist Republics;
>
> ... protection, maintenance and revitalization of a complex of military architecture at Westerplatte.[44]

---

44 "Zarządzenie Ministra Kultury i Dziedzictwa Narodowego z dnia 6 września 2016 r., w sprawie połączenia państwowych instytucji kultury—Muzeum II Wojny Światowej w Gdańsku i Muzeum Westerplatte i Wojny 1939 oraz utworzenia państwowej instytucji kultury—Muzeum II Wojny Światowej w Gdańsku," Mkidn.gov.pl, September 6, 2016, http://bip.mkidn.gov.pl/pages/dzienniki-urzedowe-mkidn/dziennik-urzedowy-2016.php.

This meant that the new museum would be almost exclusively devoted to military activities and focus mainly on the defense of Westerplatte and the battles of September 1939. In this case, most of the exhibitions, prepared for so many years and already installed in the museum, would have no reason to exist, just as most of the artifacts that were supposed to be in these exhibitions would not see the light of day. They would not fit in the military narration, defined as the depiction of only the battles in which Polish soldiers had participated, primarily in 1939. Despite the preservation of the current name, it would be a museum dedicated largely to the defense of Westerplatte and the defensive war of 1939, inserted into a gigantic building erected for much wider purposes. The whole "international" and "civilian" dimension of the museum, so fiercely attacked from the very beginning, would simply be annihilated.

We wondered how Gliński and Sellin imagined it all happening. Would the exhibition elements produced, and in some cases already installed in the building, be removed and destroyed? And what about the finished, meticulous concept of the entire museum, which required such enormous work and so much money? The only answer was that it would be thrown away. I think, as is indicated by subsequent events, that the ministers did not ask themselves these questions, did not even know how far along we were in the production and fabrication of the exhibitions. They did not have the slightest idea what an undertaking it was to create such a museum. They were not interested in our real work but simply wanted to remove me and my colleagues as soon as possible. They must have thought that it would be possible to do so before all the exhibitions were fabricated and installed. It would then be possible to make changes and announce that the Museum had been "saved for Poland." At the same time, they would present much of the work already done by us as the achievement of the new political powers. The December 1 date gave them the time they needed to prepare the legal and material framework for the merger, which involved, among other things, an inventory of our property, which was required by law and completely impossible during the ongoing construction and production of the exhibitions. However, the ministers apparently did not bother to think about such details.

The ministerial order had one positive effect for us. It could be appealed to the administrative court; the earlier minister's announcement, which was only a declaration of intent, could not. We were convinced that the actions of the minister of culture broke the law, and now we could present our arguments in a case filed in the Provincial Administrative Court. We accused the minister of culture of violating the Museums Act. First of all, as required by the law, he did not obtain the support of the Museums Council, a body comprising directors of the most important Polish museums, which is obliged to comment on any planned merger of museums. I am sure that this was a deliberate omission on the part of Minister Gliński, who expected that Polish museum professionals would not support his actions. The Museums Council had already chosen to comment on the merger in June 2016, unanimously criticizing it.[45] The minister did not take the council's opinion into account at all.

The Museums Act also stated that museum mergers may only be carried out on the condition that this step will not "harm" the functioning of the institutions and the performance of their existing tasks. We pointed out that in our case implementing the minister of culture's plan would lead to very serious chaos and delays in the decisive, final phase of construction and assembly of the exhibitions. This would be unavoidable if the Museum, the institution managing a gigantic and extremely expensive investment, was formal liquidated. It would also lead to a fundamental narrowing of the profile of the new institution (the museum of Polish arms instead of the museum of global conflict). The liquidation could mean very serious financial losses incurred by dismantling the already finished parts and changes to exhibitions, which cost over PLN 40 million (€10 million).

Another evident "harm" would be to the collection of artifacts. As I informed the court,

> Already many donors have announced the withdrawal of their artifacts, including those on permanent display, if the future of the Museum of the Second World War in its current shape is threatened. It would jeopardize the integrity of the permanent exhibitions, deprive the public of the opportunity to see unique artifacts of great historical value, and pose a threat to the artifacts themselves. Many of them require immediate conservation, which they receive at the Museum. The donors trusted specific individuals who created a specific museum, whose shape the donors knew and accepted. Their trust has been acquired over many years and can be lost instantly as a result of ill-considered institutional transformations.[46]

The city of Gdańsk also appealed to the court, indicating, among other things, that huge financial losses for the state treasury would occur if the donation of the plot was withdrawn. The third-largest complaint, however, was submitted by a constitutional officer, the ombudsman for the rights of citizens, who indicat-

---

[45] Uchwała no. 2 Rady do Spraw Muzeów, June 27, 2016.

[46] The case filed by the director of the Museum of the Second World War against the order of the Ministry of Culture and National Heritage, dated September 6, 2016, on the merger of two cultural public institutions, the Museum of the Second World War and the Museum of Westerplatte and the War of 1939, and the creation of the cultural public institution of the Museum of the Second World War in Gdańsk, September 21, 2016.

ed that the minister's order infringed on many legal provisions, including the constitution. The ombudsman, Adam Bodnar, argued that the minister's order was "inconsistent with the principle of citizens' trust in the state and law, expressed in art. 2 of the Constitution of the Republic of Poland. This constitutional principle assumes that in the lawmaking process and its application, citizens have the right to full knowledge of the premises of public authorities. This right is violated—as in the present case—when the public authority, while making a decision, does not disclose the true and actual causes of its resolution."

The ombudsman alleged that the reasons given by the minister for merging the museums "are apparent and do not correspond to the actual state of affairs.... After all, one does not create a state museum to, less than four months later, liquidate it in a merger." He indicated that the argument about optimizing the potential of two museums was untrue, since when the minister's intention was announced, "an institution called the Museum of Westerplatte and the War of 1939 did not have any potential to optimize."[47]

For the minister of culture, the complaint of the ombudsman was devastating, not only because it significantly expanded the catalog of legal violations in relation to the charges formulated by the Museum of the Second World War and the city of Gdańsk but also because it directly exposed his sinister manipulation of law and public opinion. It was not surprising then that not only I but also Adam Bodnar became the subject of increasingly aggressive attacks by Gliński and Sellin.

The "liquidation" machinery launched by Minister Gliński, however, was moving ahead despite the filed complaints. Zbigniew Wawer, a military historian whom minister Gliński appointed as his representative for the merger, arrived at the Museum. We knew him, since as a collector he had sold us uniforms when we were beginning to create the collection. This was a surprise to us, because Wawer was not affiliated with Law and Justice. His contract as the director of the Polish Army Museum had not been renewed, and he did not hide his disappointment. In a conversation with my colleague, Piotr M. Majewski, Wawer commented sarcastically that he had been captured in too many photographs with Bronisław Komorowski, the former president of Poland from the Civic Platform, to expect a renewal. However, now he had agreed to act as a "liquidator," coming to the museum in a company of lawyers. Deputy Minister Sellin told the media that it was Zbigniew Wawer who would decide when and in what form the museum would open, so it was evident that he had an important role to play.

---

47 The case filed by the ombudsman, October 20, 2016.

Half an hour before his arrival, I learned that picketers awaited Wawer at the entrance to the museum and that on the door to our office in the heart of the Old Town, there were ribbons saying, "Mr. Wawer, have some decency!" We did ask them to stop demonstrating and to remove the ribbons, which were offensive to the "liquidator." Of course, the visit was very unpleasant; we felt as if undertakers had arrived— when we were so lively and frantically busy—to talk to us about our funerals. However, the visit also felt bizarre and even grotesque, like many other actions taken against us by the ministry. The "guests" themselves admitted that they did not know how to set up the merger. They did not hide the fact that the Museum of Westerplatte was only a "shell" and that there was virtually nothing to merge from that side.

The whole operation was unprecedented. The only thing that came to their minds were analogies with the merger of large corporations. We informed them that the construction, production, and assembly of the exhibitions were underway. The detailed inventory of our property required by law was therefore practically impossible, as the situation on the construction site changed every day, bringing new elements, and part of the production was taking place at subcontractors' facilities in dozens of places outside Gdańsk. Conducting the inventory would require work to stop for at least a month, and this would amount to millions of euros in losses (including penalties for delays and perturbations resulting from disturbing the production schedule and supply for dozens of entities), for which I did not intend to be responsible. The group admitted that this might indeed be a problem. I called for the date of the merger to be moved so that it would be possible to finish the construction beforehand. I argued that the formal liquidation of the facility, which represented the largest investment in Poland in the field of culture, was madness and would lead to unimaginable chaos. Wawer, however, declined my suggestion to visit the construction site so that he could understand what he was planning to suspend. He replied that he was completely uninterested in a visit and that his mission was solely the merger. He also gallantly declared that he was not interested in my job because he did not want to leave Warsaw. He suggested, however, that he would have other important tasks entrusted to him by the government.

In addition to the lawyers, Wawer's team consisted of ten people, accountants and museologists called together from various institutions. The Museum of the Masovian Nobility in Ciechanów was particularly strongly represented, for reasons unknown to me. One member of the team refused to introduce himself and name the institution where he worked, repeating that it was "public administration in a broad sense." He obviously represented the Secret Service and might have been there to further intimidate us. In the following months, this group constantly bothered us, carrying out an inventory, examining all financial operations, and pulling employees away from urgent tasks related to construction and the exhibitions.

Information obtained from the Ministry of Culture showed that for the work of Wawer's team up to November 30, 2016, alone, Polish taxpayers paid about PLN 200,000 (€50,000).[48] Once again, the argument that joining the museums would save public funds turned out to be false. More and more money was being spent to liquidate us. We calculated that up to the fall of 2016, this already added up to more than PLN 1 million (€232,000). The liquidator himself, a few months later, became a director of the Royal Łazienki Museum in Warsaw, the most splendid classicist mansion in that city. To make space for Wawer, the previous director of that museum was dismissed, in violation of the law, before the end of his term. We so admired Zbigniew Wawer; he turned out to be a Renaissance man, not only a specialist in military history but apparently also an expert in the art and culture of the Enlightenment.

I gave Wawer official estimates, which we received from the general contractor of the building and from the exhibition fabricators, of losses that would be unavoidable if the work had to be stopped to carry out an inventory. A few days later, Minister Gliński decided to postpone the date of the merger from December 1, 2016, to February 1, 2017. Apparently he was afraid of being responsible for measurable financial losses, which, once the rule of Law and Justice ended, could even mean a criminal case. Yet again, we learned about the decision of the minister from the website of his ministry, which seemed to have become an established pathway of communication. The whole situation revealed complete incompetence on the part of the leadership of the Ministry of Culture, which could not foresee the consequences of its actions, had no knowledge of the actual situation at the Museum, and was not even interested in learning more. Gliński and Sellin, despite my repeated invitations, did not come to the Museum; they did not see the construction or the exhibitions, which in the last months of 2016 were materializing in the building at an increasing pace.

We had, however, gained another two months. In a fit of black humor, we mocked the ineptitude of the ministers who could not even hang a condemned man properly. In November, unexpectedly, events turned in our favor. The Provincial Administrative Court in Warsaw ordered the immediate suspension of all activities around the merger until the final resolution of the submitted cases[49]—a decision called for by me, the city of Gdańsk, and the ombudsman. We pointed to negative and impossible- or difficult-to-reverse effects of the merg-

---

48 Adam Leszczyński, "Koszt likwidacji Muzeum II WŚ," *Gazeta Wyborcza*, September 29, 2016.
49 Postanowienie Wojewódzkiego Sądu Administracyjnego w Warszawie, November 16, 2016.

er (including the demise of the artifact collection, changes to the exhibitions, and the loss of jobs for all employed in the Museum of the Second World War), which might later be ruled illegal. If that happened, the court would have to order a return to the previous state, which would be extremely difficult to carry out.

Minister Gliński appealed to the Supreme Administrative Court, and the matter went on, but we gained further arguments and, most of all, valuable time. It seemed we now had a chance to complete the construction and fabrication of the exhibitions and perhaps even to show them to the public, which back in April had seemed to me completely impossible. In November, in a conversation with a *New York Times* journalist, I said that the government hoped that its "genius" merger trick would allow for a rapid takeover of the Museum and its re-creation. It turned out that we resisted fiercely. Instead of the blitzkrieg on which Gliński and Sellin were counting, we gave them Stalingrad, with a fierce battle fought for every street and every house.[50] The fiasco of the ministerial offensive was increasingly noticed by the public. It was one of very few cases where it turned out to be possible to stop the Law and Justice wrecking ball, which was demolishing other institutions, including the Constitutional Court, at an impressive pace. Media widely commented on this situation, sometimes in a satirical way. I quote a fragment from the weekly humor column in an influential magazine, *Polityka*:

> The costs of the "change for better"[51] are rising, and where is one to find the money? I read that the Law and Justice government will spend around 200,000 PLN by November 30 to remove Prof. Machcewicz from the position of director of the Museum of the Second World War. As we all know, the removal of Prof. Machcewicz is one of the government's priorities and is a large-scale investment. Minister Gliński has been trying to have Prof. Machcewicz recalled since spring of this year. And to this end, he has even decided to combine the Museum of the Second World War, which Prof. Machcewicz is managing, with the Museum of Westerplatte and the War of 1939. The merger of the two institutions is being prepared by Gliński's team of specialists, and it is precisely for their work that the government will have to pay until the end of November the mentioned 200,000 PLN, and even more in the following months. Of course, if Machcewicz showed some goodwill and left, Minister Gliński would not have to combine the Museum of the Second World War with the Museum of Westerplatte. He would not have to appoint a team of specialists for this purpose, and the government could spend the money saved on, for example, cofinancing investments related to environmental protection. And the needs there are indeed enormous,

---

50 Rachel Donadio, "A Museum Becomes a Battlefield over Poland's History," *New York Times*, November 9, 2016.
51 The propagandistic term used by Law and Justice as a self-label.

since the government emphasizes that its goal is to protect various environments, especially the environment friendly to the government and the ruling party.[52]

Such texts and, above all, this inability to quickly disperse a group of historians, despite directing against them the entire state apparatus and the governmental media, had to be very frustrating. My guess is that party leader Jarosław Kaczyński must have expressed dissatisfaction with Deputy Prime Minister Gliński; the former repeatedly expressed his interest in our institution. The result was increasingly harsh language and unconcealed aggression directed toward the Museum and myself as, to my surprise, I rose to the rank of the government's public enemy number one. The lines of attack in the campaign against us also changed. After we widely circulated the catalog of the exhibitions, which proved their actual shape, it was difficult to accuse us of being a "cosmopolitan," "non-Polish" museum. In addition, the disclosure of the reviews showed that their accusations were on shaky ground and diminished their authors in the eyes of the general public; even Gliński himself must have been aware of that. As a result, the ministerial gears shifted to discrediting the Museum and myself through allegations of spending irregularities.

The minister sent ten auditors to the Museum, who for a few months reviewed piles of documents, looking for any possible shortcomings. At times, they completely paralyzed the work of the financial and investment departments, whose employees, instead of supporting the gigantic project, had to find thousands of documents, sometimes from many years ago, in immediate response to dozens of formalized questions. This meant that during the most important months of completing the investment and preparing for operations, we were tormented by completely unnecessary actions aimed at discrediting the Museum and closing the institution before it even opened.

The head of the audit team was a longtime director of the finance department of the Ministry of Culture and National Heritage, Wojciech Kwiatkowski. He had supervised our investment on behalf of the ministry from the very beginning, had made numerous decisions about it, and in fact was responsible for it. Now, without a blink, he accused us of fundamental irregularities in matters to which he had previously expressed (written) consent. Finally, I understood the art of survival and how it was possible for Director Kwiatkowski to hold a position over several years, outliving many ministers and numerous political parties in power.

---

52 Sławomir Mizerski, "Oczyszczanie środowiska," *Polityka*, October 5, 2016.

It was clear to me from the very beginning that the audit was an excuse for obtaining materials that could be used against us, but I accepted it with relative calm. First of all, I knew that any allegations that the minister presented would be untrue. The Museum had just undergone a months-long, meticulous examination by the Office of the Auditor-General, a public institution independent from the government, which concerned exactly the same issues now being scrutinized by Minister Gliński's subordinates. I referred the latter and all interested journalists to the Office of the Auditor-General's final report, dated December 2015, which confirmed no fault with the "reliability, frugality, and purposefulness of the museum's operation."[53]

All the points of reference had now reversed: the subject of a positive evaluation by an independent constitutional body appointed to control public entities was now being questioned by ministerial officials. The best troops were sent to the museum front. The minister's accusations were published in the right-wing newspaper *Gazeta Polska* under the dramatic headline "Most Overpaid Museum in Poland."[54] The journalist's role was to rewrite fragments of the ministry's most recent financial review, passed to him by the ministry, but I do not think that he actually understood it all. Over the following months, ministerial officials and Gliński and Sellin tried to present me as a man guilty of serious irregularities, who at all costs wanted to avoid responsibility, and that this was what the entire struggle for the Museum's survival was about. Here is a sample of what Deputy Minister Sellin himself submitted to Jerzy Halbersztadt, vice president of the Polish chapter of the International Council of Museums, in response to its defense of the Museum, which had been signed by hundreds of Polish museum professionals. According to the minister, the conflict "in no way concerns, as suggested by the current director of the Museum, the content of the main exhibition, but results from irregularities in managing investments, confirmed, among others, by the Office of the Auditor-General."[55] Sellin had to be fully aware that the Auditor-General's report had contained a completely different assessment of the investment for which I was responsible, but even more interesting was the fact that he pretended that there were no attacks on the shape of the exhibitions, especially not from him.

A frontal attack on me for "irregularities" was personally carried out by Minister of Culture Gliński and supported by Law and Justice senators during a meet-

---

53 Uchwała Zespołu Orzekającego Komisji Rozstrzygającej w Najwyższej Izby Kontroli, December 15, 2015.
54 Piotr Nisztor, "Najbardziej przepłacone muzeum w Polsce," *Gazeta Polska*, December 21, 2016.
55 Jarosław Sellin to Jerzy Halbersztadt, March 7, 2017.

ing of the Senate Committee on Culture and Mass Media. He presented in detail the eleven charges against me, which at times sounded surreal. "The auditors found the museum's management responsible for designing a faulty structure that is not resistant to weather conditions." The minister of culture read from a script, also repeating the accusation about locating the museum in the "port basin," which had no relation to the facts. Senator Czesław Ryszka expressed his concern about "the safety of collections and the safety of visitors, because the exhibition hall is 17 m underground, and there are leaks."

I answered all the allegations related to the engineering. I assured the senators that the artifacts and visitors were safe, although at that moment I had the impression of participating in the theater of the absurd. Jacek Taylor, in the time of the Communist dictatorship a human rights advocate and now a member of the Museums Trust Council, explained to the senators the geological conditions and the nature of construction in Gdańsk, passionately defending the Museum: "You, Ladies and Gentlemen, have talked here many times about the location, that it is somebody's fault. Well, I'm quick to explain that the biggest brick-walled church in all of Europe is located nearby, the St. Mary's Basilica, which has been standing on stilts for 500 years, like many buildings in the vicinity of the almost-completed Museum of the Second World War."

I also invoked the favorable results of the examination carried out by the Office of the Auditor-General. Minister Gliński questioned them and suggested that the audit was "fixed," although he did not explain by whom and how: "We also have information about why there were not more charges. This is interesting and it will be checked." The minister of culture also questioned the intentions of the ombudsman, accusing him of political motives for filing a complaint against the merger of the museums.

Then, there was a litany of the usual personal allegations against me. "You, Sir, have launched a huge, monstrous, negative, black campaign directed at me and my institution. And this is quite obvious to me," Gliński accused. This conspiratorial vision of reality, coming from the deputy prime minister, who as a sociologist had studied civil society, surprised me. He did not want to accept that people could, on their own initiative, join in the defense of values they consider important.

"I did not launch any campaigns," I replied to the deputy prime minister. "Rather, citizens and the media are interested in the matter. Many milieus express indignation at your actions. They have the right to do so. And I really do not inspire anything; nor do I have to inspire. Social life is governed by certain autonomous rules. And I am not the author of any plot, against either you, Minister, or the ministry."

There were also accusations from the minister that I was a politician and even that I wanted to take his place. The final exchange between Gliński and me had an overtly grotesque aftertaste:

> Minister ... sorry, Director, maybe a minister in the future.
>
> No, Minister. The function of a minister would bore me; so much organizational and administrative work ... So, I will not be running for office. I plan to write books.[56]

All this took more than four hours. Gliński and I sat facing each other, and at the end it was a verbal confrontation between the two of us with the whole room listening. In every word and gesture of the deputy prime minister, I sensed his rage and frustration that the struggle against the Museum, despite the disproportion of forces, had unexpectedly proven to be such a long "war of attrition." The inability to immediately take control of the Museum provoked more and more hostile steps, which were also increasing the psychological pressure.

Gliński and Sellin declared that the decision of the Provincial Administrative Court to suspend the museums' merger was not legally binding and that the ministry would not stop the process. It would be an ostentatious violation of the law, but it had already happened in Poland on many occasions during the previous year. The ombudsman for the rights of citizens called on the minister of culture to abide by the law and refrain from continuing the process. Nothing indicated that he would. On the order of the minister, Wawer tried to continue, but in the face of the court's decision, I refused to provide him with documents. I was even threatened with criminal liability for alleged irregularities. Just before Christmas, the Ministry of Culture sent an ominous message, which announced that "some of the opinions assessed in the course of the audit of the Museum management's operations exceed the definition of actions that are not legally binding and harmful not only for the Museum but also for the Treasury." Unspecified but "adequate actions" were to be taken.[57]

There was also, simply, harassment against me personally, perhaps with the hope that I would have enough and give up. At the end of December, I received an e-mail invitation to participate in a public discussion about the memory of the Second World War. Andrzej Nowak was supposed to be the second guest. It was

---

56 Senat RP. Zapis stenograficzny. Posiedzenie Komisji Kultury i Środków Przekazu (31), October 26, 2016. IX kadencja, www.senat.gov.pl.

57 "Komunikat MKiDN w sprawie połączenia Muzeum II Wojny Światowej i Muzeum Westerplatte i Wojny 1939," Mkidn.gov.pl, December 21, 2016, http://www.mkidn.gov.pl/pages/posts/komunikat-mkidn-w-sprawie-polaczenia-muzeum-ii-wojny-swiatowej-i-muzeum-westerplatte-i-wojny-1939 – 6958.php?p=10.

organized by the recently established Center for Research on Totalitarianisms. I was surprised because everyone knew that I was blacklisted, but I accepted the invitation. However, after a few days, the deputy director of the center called and informed me that he must withdraw the invitation. The Ministry of Culture opposed it and stated that my participation in the discussion would have adverse consequences for the center. I made the issue public because it reminded me of the practices of the Communist regime.[58]

Then, Łukasz Michalski (my colleague and at one time an employee of the IPN Public Education Office, which I created) joined the fight in the role of an online troll. He was an advisor to Deputy Minister Magdalena Gawin, who was responsible for the center. Michalski argued on Facebook that the invitation had only been "preliminary"; he organized many conferences, and it was allegedly standard practice to withdraw some invitations to sit on a panel. And in general, Michalski argued, questioning the withdrawal was a disgusting provocation, deliberately arranged by me to attack the minister of culture. Apparently appreciating his merits, a few months later the latter appointed Michalski as director of the State Publishing Institute.

I found it interesting to observe the behaviors of many people and to compare them with the behaviors that I studied and described as a historian dealing with the times of the Polish People's Republic. When I attended public events, many well-known people in managerial positions in the public sector avoided me. Some of the braver came up to say hello but looked around carefully, and when I started talking, they were gone, and I was left talking to myself. The epidemic of fear was contagious, and many people raised in free Poland turned out to be completely unprepared to face unexpected challenges.

Many people voiced support and solidarity, however, which allowed me to survive the worst moments. The Forum of Researchers of Contemporary History expressed solidarity with the Museum. It comprised more than two hundred historians from all over the country, who convened in December in Warsaw under the aegis of the Polish Historical Society.[59] For me, most touching was the support of people with whom I was not professionally associated and whose support I did not expect. Krystyna Zachwatowicz sent a letter to Minister Gliński, thanking him for attending the funeral of her husband, acclaimed film director Andrzej Wajda. "As you well know," wrote the widow of the great director,

---

58 Paweł Kośmiński, "Dyrektor Muzeum II Wojny niemile widziany na debacie historycznej? Wracają czasy dyrektyw z KC," *Gazeta Wyborcza*, January 3, 2017.
59 Adam Leszczyński, "Alarmowy zjazd historyków," *Gazeta Wyborcza*, September 27, 2016.

we were both from the beginning associated with the creation of the Museum of the Second World War in Gdańsk. Andrzej Wajda was a judge in the competition for the design of the permanent exhibitions. You certainly know about our protest against the changes you have announced in the scenario of the exhibitions and the decision to combine the Museum of the Second World War with the nonexistent Museum of Westerplatte. I want to ask you in this letter to desist from these intentions. In that way, you will best honor the memory of Andrzej Wajda.[60]

I also received letters from donors I had never met in person. Here is an excerpt from one of these letters:

I am a lawyer by education and have worked as a legal advisor throughout my professional life. However, I became interested in history in primary school, and this interest has accompanied me throughout my whole life... . Maybe because this war has left a mark on me that I have felt all my life. My father set out on September 4, 1939, from the Poznań Citadel along with the Seventh Telegraph Company of the Poznań Army. I was 81-days old then and never saw my father again. Interned in Hungary, he survived Auschwitz and Neuengamme camps, only to die a few days before the end of the war on the infamous "ships of death" in the Bay of Lübeck. As a result of the war, I have never experienced fatherly love, care, or support. It was difficult to see friends walking with their fathers for a match or on a trip. That is why I feel empowered, as probably many Poles do, to speak about what a certain sociologist [Gliński] has created around the Museum... . No sensible person will say that the war has any face other than death, disability, and destruction. All these stories about heroism, patriotism, or character building are mostly later inventions of propagandists who spent the war in safety or, as at present, watched on television, or learned from so-called historical re-enactments or paint-ball instructors. One must always and everywhere say and show that every war is the greatest evil that humanity can encounter. And I say that the exhibition reached this goal one hundred percent. Professor, I've written too much, but I wanted to express that there are people who think like you and your team. And there are people who strongly support the vision of the Museum contained in the catalog.[61]

Here is another excerpt from a letter with a slightly different tone, from a donor to whom we had sent the exhibition catalog. Piotr Kwieciński wrote,

I would like to express my appreciation to you, the Academic Advisory Committee, and the whole team for the concept and shape of the exhibitions under construction. In my opinion, just as Poland's membership in the European Union and NATO brings our country economically and militarily back to the democratic West, so the idea of entering the fate of our country into a comprehensive context of the Second World War places us again among the nations of the world and not, as has been the case so far, on the world's margins. The value of this is priceless. Undermining it is a crime.

---

60 "Jak uczcić Wajdę. Wdowa po reżyserze pisze do prezydenta Andrzeja Dudy i ministra kultury Piotra Glińskiego," *Gazeta Wyborcza*, November 23, 2016.
61 Letter from Wojciech Rybarczyk, July 20, 2016.

Dear Professor, I watch, listen, and read as the small-minded people, using the language of Berman, Moczar, Gomułka, and Urban,⁶² for their own low reasons try to destroy your work, Sir, and the work of your team. I am writing this letter for one purpose only: that the professor would know that an ordinary, everyman like me sees and appreciates the work done on the creation of the Museum of the Second World War. He sees the scale of lies and slander against the creators and the institution so unique in Europe. That this everyman deeply believes that this work cannot be destroyed. Not by anybody ever.⁶³

On the streets, especially in Gdańsk, people unknown to me approached me and asked me to persist in defending the Museum. They assured me that they were with me and with all our team. I was aware that the Museum had become a symbol of values that were important not only to me and my colleagues but also to a large number of Poles. Without setting out to do so, we were on the front line of the struggle to preserve the autonomy of culture and history in Poland, as well as the rule of law itself. For me it was more and more obvious that we could not give up and had to open the Museum against all odds.

---

62 Well-known Polish Communists, generally associated in public memory with propaganda and Orwellian newspeak.
63 Letter from Piotr Kwieciński, July 17, 2016.

The Finale

It was not easy to keep going, and not only because of the uncertainty of our situation and the open hostility of the Ministry of Culture, which had been going on for so many months. The lack of money and staff needed to open and run a large museum posed a serious problem. The ministry awarded us an operating budget for 2017 of approximately PLN 11 million (€2.55 million), similar to the funding that we had received in 2016. I had applied, however, for an amount almost twice as high; this was a considerable difference. Up to this point we had rented a dozen or so rooms in a temporary office, but after the opening of the Museum, we would incur a new set of expenses related to its operation and maintenance: energy, water, cleaning, and security. That would add up, according to our estimates, to almost PLN 10 million (€2.317 million), and it was unclear how we would pay salaries.

At the end of 2016, the Museum employed about sixty people, including ten engineers whose salaries were paid not from our annual budget but from the funds of the Multi-Year Government Program, which would expire on the completion of construction. We assumed that in order to function normally after opening, we would need about 120 employees, and this number was accepted at the outset of the Ministry of Culture's investment. For comparison, in 2016 the Museum of the History of Polish Jews in Warsaw employed about 150 people and had an annual budget of PLN 35 million (€8.109 million). We estimated that we would need to hire a large group of visitor services employees at least two months before the planned opening. They needed to be trained and prepared for their duties before thousands of people started visiting the Museum. In our financial circumstances, this was completely impossible.

In letter after letter addressed to Minister Gliński, I raised alarms that the budget granted to us essentially hindered the opening of the Museum and certainly prevented its normal functioning. I argued that the Polish taxpayer had allocated a great deal of money to its construction, so if the Museum did not open, the investment would be thrown out the window. I argued that the Museum should carry out extensive educational activities, since that was its mission, and that the new building would offer excellent conditions for such activities. However, new employees were needed; an understaffed education department would not be able to support the increased requirements. My appeals had no effect; they were almost counterproductive. The minister of culture made further cuts in our budget: he cut PLN 120,000 (almost €30,000) first, at the end of 2016, and then another PLN 420,000 (approximately €100,000) the following January. There was no explanation for the cuts. It was clear to me that this was another means of fighting us and a way to prevent the opening of the Museum. This, however, only amplified our determination.

We made a decision at the end of 2016 that if we could survive even a few months, we would open the Museum at any cost, knowing that the money might later run out. Starting in January 2017, we suspended all educational, research, and publishing activities to maximize our funds. All funding was devoted only to those activities that led to the opening of the Museum and its operation at the most elementary level for several months. We also decided to employ a few additional people to staff the cash registers, without whom we could not welcome visitors. All other tasks were to be conducted by the current team, even though, staff-wise, it was completely inadequate to meet the needs of the Museum after its opening. We had to give up hiring our own guides; instead we trained external guides already operating in Gdańsk. By the time of my departure from the Museum, the external guide team already counted about sixty people, who were prepared to show visitors around the exhibitions in many languages, including Chinese. We estimated that total revenues would give us a chance to survive until the fall of 2017. This was as far ahead as we could see. I did not think I had a chance of surviving that long as director. It was obvious that the new director, appointed by Minister Gliński, would receive additional funds. The fight was only about the possibility of opening the Museum and presenting our work to the public. It would be a huge victory, which just a few months earlier had seemed to me completely beyond our reach.

The situation was made even more complicated by the statements of Gliński and Sellin that the museums would definitely be merged on February 1, 2017, and that they were not bound by the court's decision. This would be pure lawlessness, and after consultation with Jacek Taylor (who had a long record of confronting the authoritarian practices of the Communist authorities), we felt that we could not agree to it. Of course, I was going to respect all court decisions, but I expected the same from the minister of culture. If the minister's people wanted to enter the Museum on February 1 and remove me from the office, they would have to involve the police. It would happen in public, in front of television crews, and it would create an unimaginable scandal, but I could not rule out that it might actually happen.

In December, we learned that the Supreme Administrative Court would consider the minister's complaint on January 24, although generally the waiting period was much longer. On the one hand, this would provide a final judgment before February 1, thereby avoiding the most dramatic scenario. I was certain that, in any event, I would be removed from office under the pretext of alleged irregularities "disclosed" during the ministerial audit, but I was hoping that this would not happen immediately. On the other hand, such a fast mode of operation by the court meant that we really had little time left. If the court ruled against us, February 1 would be a definite end. Therefore, we decided to show audiences a preview of the still unfinished exhibitions. We were only few weeks shy of completion, but we were not sure that we, in fact, had these few weeks. On January 23, the day before the court ruling, we invited to the Museum historians, museum professionals, journalists, veterans, and donors—a total of several hundred people. We announced that on Saturday, January 28, and Sunday, January 29, we would hold an open house for all interested parties. The entrance tickets were available online, and in just a few hours had all been distributed.

The last days were complete madness for both the exhibition fabricators, who carried out the assembly work at an unimaginable pace, and the Museum employees. During the night of January 22–23, we installed artifacts in display cases. We estimated that the first guests saw about 70 percent of the exhibitions, and those who came a few days later, during the open house, saw even more.

The *Gazeta Wyborcza* began its report about the Museum, "'*Good morning, my name is Paweł Machcewicz; I am the director of the Museum of the Second World War.*' After these words, several hundred people gave him a standing ovation. It was an expression of support for Machcewicz's efforts to maintain the independence of the institution."[1] For me it was a moment of great satisfaction and personal triumph, one of the most important moments in my life. I had not thought it possible to survive until the exhibitions, even unfinished, could be shown to the public.

The visitors' reactions were largely enthusiastic, as were those of the general public, veterans, and donors, as well as the world's greatest historians and museum professionals. Their debate, after seeing the exhibition, was the first attempt to evaluate it. Timothy Snyder said,

> There are many Second World War museums in the world, but the narrative of the exhibition in Gdańsk completely changes the perception of the war; such a museum is a civilizational achievement. The exhibition is very versatile because it talks about the fate of civilians around the world, but at the same time it is also very Polish; it does not lack any major events related to the wartime history of Poland. Polish politicians and researchers often complain that the world does not understand them. If this is the case, then an exhibition at the Museum of the Second World War is an opportunity for them to finally be understood. Every change in it will result in the destruction of a coherent narrative. The Museum of the Second World War cannot be more military or talk more about the war in 1939 or focus more on the fate of Poles. Due to the fact that the exhibition captures the war in a

---

[1] Krzysztof Katka and Emilia Stawikowska, "Bitwa o drugą wojnę. Muzeum w Gdańsku pokazało wystawę, nikt z PiS nie przyszedł," *Gazeta Wyborcza*, January 24, 2017.

global way, it puts Poland and Gdańsk at the center of the debate on the history of the Second World War.

For me, the assessment of Sara Bloomfield, director of the United States Holocaust Memorial Museum in Washington, DC—the first narrative historical museum in the world, which created a pattern for many other institutions in this area—was especially important.

> There are many museums in Poland devoted to particular aspects of the Second World War, but there is no one that talks about it comprehensively. The architectural shape itself and the location of the museum appeal to visitors. People do not know much about the history of the Second World War. For example, Americans think that they alone won the war, and their knowledge about the theater of war is limited to the fighting in the Pacific. That is why a place such as the Museum of the Second World War, which will tell them about the history of the war very comprehensively, was needed.[2]

The reactions of foreigners, including journalists, proved that the idea of creating a comprehensive story about the war that incorporated Polish history worked. That it was being realized in a museum that the Polish government had fought with all its force, while constantly declaring that we were a nation misunderstood, disregarded, and accused of complicity in the Holocaust and other evil deeds, was a bitter paradox.

It was a day of great joy for the whole team that had worked on creating the Museum for so many years under increasingly dramatic circumstances. The next day, however, the situation changed completely. During the lunch for members of the Academic Advisory Committee, I received a phone call saying that the Supreme Administrative Court had sent the Museum case back for reconsideration to the Provincial Administrative Court, perceiving its earlier decisions as flawed. At the same time, it overturned the suspension of the museums' merger. This meant that on February 1, just a few days after the exhibition was presented to the public, the current Museum of the Second World War would be formally closed down. We did not believe that the lower court could reexamine the case in the next few days.

Two days later, another blow clearly showed the surgical precision with which the ruling camp was destroying people and institutions that it regarded as enemies. It started on the morning of January 26 with a question that a Law and Justice MP, Barbara Dziuk, asked the deputy minister of culture in

---

2 Emilia Stawikowska, "Wrażenia gości po wizycie w Muzeum II Wojny. Timothy Snyder: *To osiągnięcie cywilizacyjne*," *Gazeta Wyborcza*, January 24, 2017.

the Sejm: "Two questions for the minister: Is it true that the director of the Museum of the Second World War in Gdańsk built a hotel complex with luxury apartments? And the second question: Is it true that the director purchased illegal household appliances contrary to the provisions on public procurement? If so, what was the cost?"[3]

Minister Sellin immediately came to the podium. The model state official, he did not let himself be surprised by this detailed question and gave an exhaustive answer.

> The museum, which the MP was particularly interested in—and specifically in the matter related to the construction of this museum—the Museum of the Second World War in Gdańsk was established in 2008 and was to open in 2014. It is still not open today. Initially, 360 million PLN was planned for the construction of this museum in a Multi-Year Government Program. This amount was changed to 450 million PLN. It is, for the time being, the most expensive museum in the history of Poland, although not yet finished, unopened, and unfinished. And responding specifically to the lady's question about, as you put it, building during the construction of this museum a hotel complex with luxurious apartments, I would like to inform you: indeed, on the occasion of construction of this museum, eight residential units were built along with access routes located on the first floor and in parts of the second floor of the second museum building... . These rooms are equipped with completely furnished bathrooms, bedrooms, seating sets, radio and television equipment. Each of the apartments consists of a kitchen, a dining room, a living room, two bathrooms, two bedrooms, a dressing room, and they are also equipped with household appliances, as well as radio and television equipment.[4]

Sellin also read extensive passages from the postaudit ministry protocol, which by happy fate he had with him. A moment later, we received a phone call from the Gdańsk branch of the state television network, which wanted to film the hotel complex. They were not interested in the exhibitions. State television, unlike commercial and many foreign media, had not reported on the preview of the Museum. A twenty-something journalist asked me questions from a page of handwritten notes. His questions insinuated that the apartments were built for me and my colleagues and that their equipment cost a dozen or so million zlotys. I denied both allegations, saying that such an amount had been spent on finish-

---

3 "Wypowiedzi na posiedzeniach Sejmu. Posiedzenie nr 34 w dniu 26–01–2017. 14 punkt porządku dziennego: Pytania w sprawach bieżących. Poseł Barbara Dziuk," www.sejm.gov.pl.

4 Wypowiedzi na posiedzeniach Sejmu. Posiedzenie nr 34 w dniu 26–01–2017, 14 punkt porządku dziennego: Pytania w sprawach bieżących. Sekretarz Stanu w Ministerstwie Kultury i Dziedzictwa Narodowego Jarosław Sellin, Sejm.gov.pl, January 26, 2017, http://www.sejm.gov.pl/sejm8.nsf/wypowiedz.xsp?posiedzenie=34&dzien=4&wyp=31&symbol=WYPOWIEDZ_PYTANIE&nr=324&pytID=C04EF92C7D6CD6C8C12580B4004CD24C.

ing and furnishing the entire building, with an area of several tens of thousands of square meters. This information was not broadcast. I asked the reporter what was on the piece of paper from which he read the questions. He admitted with astonishing honesty that these were "issues" to be addressed in the broadcast that he had received from the Ministry of Culture.

The cameraman was taking close-up shots of door handles, shower stalls, and showers. It reminded me of materials from the 1980s, documenting the detention of activists of the Solidarity underground by the Communist Security Service. The camera cruelly showed the dollars found in their apartments, as well as foreign alcohol and cigarettes, supposedly thereby revealing the true face of the opposition activists. That evening, the state news showed a report of over four minutes under the shocking title "Hotel in the Museum of the Second World War."

Gliński and Sellin apparently could not stand the enthusiastic reactions after the preview of the exhibition and wanted to spoil the atmosphere as much as possible before the upcoming open house. They hoped that public attention would focus on the "luxury director's apartments" rather than the exhibitions. In reality, the apartments were not luxurious, merely of a standard that would allow them to be rented for the benefit of the Museum. From the beginning they were part of the building's design, which had such an attractive location that rentals could bring in several hundred thousand zlotys a year. In addition, members of the board of trustees and the Academic Advisory Committee could use them, as could all other guests of the Museum, which in turn would save a lot of money on hotel rentals. Many other institutions in Poland have similar guest rooms, and they stir up no emotion or interest.

The already-mentioned reduction of our budget by several hundred thousand zlotys was an additional "gift" from the ministers a few days before the open house—a small token of revenge for showing the exhibitions to visitors. As soon as the matter was made public, the money was restored to us, also without explanation. This proved that our tactic of going to the media had been effective.

During its two open days, over 3,000 people visited the Museum. There were long queues of people who had not managed to book tickets online. We tried to let everyone in, realizing that this could be the last opportunity to show the exhibitions as we had created them. I spent many hours on Saturday and Sunday trying to remember everything about the exhibitions. In a sense, it was a goodbye for me; there were only three days left until February 1. Once again, sure that I was doing it for the last time, I packed and emptied my office. Zbigniew Wawer's "liquidators" ended their activities, and the ministry instructed the chief accountant to guarantee funds for the salaries of the new director and two other people, the "initiative group," which would take over the Museum the next day. So it was the real end, which I faced with sadness, of course, but also to some extent with a sense of relief that the war of nerves would end and my life would return to normal. The most important thing was that at least several thousand people had managed to see the exhibitions.

I took time to thank the museum team for the years of work; I argued that it was worth the sacrifice, to create a museum and bring the audiences to the show, despite everything that had happened over the last year. I even got a farewell present, a tennis racket. On Tuesday, January 31, we agreed to take the last photo with the whole team in front of the Museum building on Bartoszewski Square. I arrived in a rather awful mood a few minutes before noon and joined dozens of people who were already there, getting ready for the photo. At that very moment a journalist from an independent private TV channel who was preparing a program about the Museum called. The information he gave me was so amazing that I took a long while to believe it. On the previous day, in the late afternoon, the Provincial Administrative Court in Warsaw had reexamined the Museum's case and upheld its earlier decision, stopping the merger of museums. Information about the decision appeared on the court's website only the next day. We took the planned photo, but in completely different moods: instead of a funeral, there was an explosion of euphoria as we threw our hats into the air. Yet the emotional roller coaster was becoming harder and harder to endure.

In justifying its decision the court referred to the issue raised earlier by the Supreme Administrative Court, which would ultimately prove decisive for the fate of the Museum. The Provincial Administrative Court took the position that the decision of the minister of culture about the merger of museums was subject to judicial control and was not merely an "act of internal management." "In the opinion of the Court," Judge Izabela Ostrowska argued, "cultural institutions such as museums are not organizationally subordinate to the minister of culture and national heritage, because the minister exercises only foundational supervision over them, strictly defined as to the competences and resources in the Museum Act... . It should also be emphasized that any doubts, emerging in practice, to the admissibility of the court-administrative proceedings, should always be interpreted in favor of the right to the court."[5]

Along with the new decision of the court, I again had an opportunity to open the Museum to the public with completed exhibitions. In January, only a few thousand people had been able see it, and the fabrication was still underway.

---

5 "Postanowienie Wojewódzkiego Sądu Administracyjnego w Warszawie," January 30, 2017, bip.warszawa.was.gov.pl.

The minister of culture again appealed to the Supreme Administrative Court, but we assumed that it would take at least a few weeks to resolve, and that was all we needed. We finished the exhibitions at a feverish pace; we installed the missing artifacts in the cases and tested the operation of multimedia stations. Under normal circumstances, this would have taken three months. We did not have so much time. At the same time, technical approval of the building took place, without which it was impossible to take over the building from the general contractor. Tests of the monitoring system were also a challenge; without them we could not make the museum building accessible to the public, since it housed thousands of priceless artifacts. Every day the exhibitions were visited by dozens of people: groups of historians and museologists from various cities and television crews from around the world. Everyone was aware that this state of suspension would not last long.

"Even if Minister Gliński carries out the liquidation of the Museum of the Second World War, the question of collision between different models of Polishness will remain open," wrote Piotr Kosiewski in *Tygodnik Powszechny*. It was one of the most insightful reviews of the exhibitions, written before the Museum was opened to the public on a regular basis. Kosiewski understood that the exhibitions offered a reading of the war and Polish experience that went far beyond the military-insurrection model so present in Polish tradition and reinforced with such determination by the Polish political Right. This model rejected both a confrontation with less obvious themes and a vision of Polishness interconnected with the experiences of other nations.

> The Law and Justice model of Polishness is not the only one possible, and as the example of the Gdańsk Museum shows, we are dealing with the collision of different models. Polishness, conscious and strong, which is able to confront its past and talk about it. Polishness that can realistically and even critically look at itself, because it has the ability to realistically assess threats and draw conclusions from them. With another Polishness, constantly assuring itself and others of its greatness, "rising up from her knees," but in fact uncertain, still scared, afraid that in a moment Poles will cease to exist. Polishness afraid of intellectual courage and critical thinking, although this attitude is a betrayal of one's own national culture, in which the reflection on faults, errors or offenses has always occupied a very important place.[6]

One day before the public opening of the Museum, one of the guides we were training wrote about his impressions on Facebook. I quote a fragment of this very long entry because it is one of the most moving reviews of the exhibitions, which also reflects the emotions of that moment. Łukasz Darski wrote,

---

6 Piotr Kosiewski, "Wystawa zbójecka," *Tygodnik Powszechny*, February 27, 2017.

I have not started working there yet, and already I do not like this museum. Because I will have to cry over every visit in this place. Just like after every visit to Stutthof [concentration camp], I experience a two-day down and a feeling of enormous pain. Nevertheless, I will go there as often as possible. I consider it my duty to my children, grandchildren, and all the inhabitants of the world who come in contact or encounter the nightmare of what people can do to other people. A nightmare systematically devised and carried out in a calculated manner. Enforced, in a more or less organized manner, but always with an iron consistency. As a result of the actions of Soviet Russia and Nazi Germany, over twelve years, 14 million civilians were murdered. Civilians. Not soldiers. Murdered, not killed. Some were starved to death; others were shot with a bullet to the base of the skull; yet others were gassed or cut down by machine-gun fire from soldiers, planes, tanks; more were burnt in their own barns; still others were forced to work so hard that they died.

This is the first museum talking about the Second World War in such a devastating way. Orderly and systematized. Presenting mainly the perspective of civilians: victims, torturers, and bystanders. In the pictures, the executioner is often in a uniform of some military formation. But it is always a civilian under a uniform who decides to rape, plunder, murder. Or sometimes he evades the decision to oppose the order. Bystanders often accompany him. With mouths open. With a smile plastered on the face. The story of a total war. War of annihilation. THIS IS TERRIBLE . ..

Why the hell am I inviting you there? Why treat yourself to such a terrible experience? For three reasons. First: to see a small section of the museum, called Resistance. The only section that gives unambiguous hope. It talks about the beautiful and painful efforts of millions of people, who resisted the horrors brought on by the Russians and Germans. You can spend over two hours there. Among the chaos of feelings bursting in me like grenades, I found peace. I had to force my mind and my will to stay there. Hold on to it tightly, like to a tree trunk tossed by floods. This section gives me hope. It restores faith in a human being.

Reason two: to encourage you to think and talk about this damn hard subject. We must talk, to be able to oppose it. To react in time, to harden the will, to soften the heart. To respond to evil in an adequate and fast manner. To save as many helpless victims as possible; it is best to save them before they become so.

Reason three: to defend the museum against the temptations of politicians who, blinded by their littleness, want to destroy this work. A work of human genius. A complicated structure of the idea, the architecture, the design, the knowledge, and the emotions. Many people have labored over the years to create this Museum.

We have the right to speak out loud about the wrongs we, Poles, have suffered for decades, of the nightmare of war and communism. However, one can do it in a variety of ways. The museum team has created a platform that allows everyone to participate in a dialogue about evil. Executioners and victims. Those who have suffered the hardest and those who have suffered the least. We all have wounds and burdens that open when we touch difficult subjects. Only respect for the other party enables agreement. The Museum of the Second World War in Gdańsk opens its doors to everyone. Come to this Museum as soon as possible. Its existence in its present form is seriously endangered. We have just over a month left. Then the grinders of a political mill can wipe it away. If, after visiting the museum, you feel and think as I do, be ready to go out on the streets in defense of the Museum.

Finally, on March 23, 2017, came the crowning achievement of many years of work by me and hundreds of other people. I invited veterans, prisoners of concentration camps, and donors to the opening. I also invited the minister of culture, but I knew well that he would not appear in the Museum as long as I was there. Therefore, I did not invite any other politicians, because I realized that only those connected with the opposition would come, and thus the Museum opening would take on a political character, which I wanted to avoid. I opened the Museum by leading in, as the first visitor, the ninety-six-year-old Professor Joanna Muszkowska-Penson, a courier of the underground Union of Armed Struggle, a prisoner of Pawiak prison and Ravensbrück concentration camp, a Solidarity activist, and a doctor who assisted opposition activists during martial law in the 1980s. She gave the Museum her letters from the concentration camp. In recent years, she had become a guardian angel and a symbol of the Museum, being with us in the most difficult moments.

The opening had an intimate character, quite different from the opening of the Museum of the History of Polish Jews or the European Solidarity Center. There were no state officials or official speeches. Everyone was perfectly aware that this reflected our situation: a museum opened against the will of its own country's authorities, who tried every means to prevent this moment and thwart our efforts to bring the work to an end and show the exhibitions to the public. Veterans, ex-prisoners of the war, and donors—the main heroes of the day—were part of the social movement that defended the Museum. The donors were also its cocreators, since the family souvenirs that they provided now became part of our story of war. Journalists talked to our guests, mostly very old people who did not hide their satisfaction and emotions. For them, the exhibitions reflected their lives.

In the evening on the opening day, the minister of culture was a guest on a TV news program broadcast by the independent Polsat News. Asked by the host why he was not in Gdańsk that day and for his comments on the fact that the veterans and donors were delighted with the exhibition, Gliński answered with a dismissive question: "Do you think that donors and veterans have any basis for comparison with other museums?" Then he attacked the Museum: "The problem is that it is a 100 percent state museum. I am responsible for cultural policy, and this is a museum that is a subject to the supervision of my ministry. Unfortunately, I have been unable to implement this supervision for over a year. For several reasons, among other things, due to the very strong and aggressive media campaign." He then went on to attack the courts, claiming, "The Provincial Administrative Court has not considered this complaint for many months, which prevents me from performing my constitutional obligations, i.e. the merger of two cultural institutions." And then, in one phrase, he summarized the work of courts: "courts on call." This was probably a reference to the quick response to the merger that the Provincial Administrative Court had issued at the end of January, which unexpectedly postponed the liquidation and gave us invaluable time to open the Museum. There was also a suggestion in this that I was manipulating the court, which I can only consider as another manifestation of the conspiracy theories of the deputy prime minister and indirectly also a recognition of my supposed influence.[7]

From the first hours of the opening, there were crowds in the Museum, even though it was not yet tourist season. On the first weekend, we had a real siege. A lot of people, including those who had come that day from other cities, could not get in because we ran out of tickets (for safety reasons, we could have no more than seven hundred people in the Museum at the same time). During the first two weeks the exhibitions were visited by about 20,000 people. The cashiers barely managed. We did not have the opportunity to hire additional people and extend the opening hours of the Museum, which would have been a natural and even necessary step in the face of such a multitude of visitors. During those two weeks, between the opening of the Museum and my removal from it, I spent several hours every day in the exhibitions, guiding historians, museum professionals, journalists, and sometimes great individuals such as Lech Wałęsa. Donald Tusk visited the Museum in April, when I was no longer there. Along the way, I had the opportunity to observe the reactions of other visitors; some of them approached me, thanking me for the Museum and for opening it despite all adversity. At such moments I had the feeling that it was all worth it.

At the same time, it was obvious to me that this situation was temporary. It was also clear to the public. As Mikołaj Chrzan wrote, "We have an impressive red pyramid over the Radunia Canal, a new symbol of Gdańsk. The exhibition is already open, reminding us of the apocalypse of the Second World War. Soon we will find out: Will this facility fall, despite the heroic defense—just like the nearby Westerplatte? Or will it, perhaps, be able to defend itself?"[8] The title of this text, published in *Gazeta Wyborcza*, was also symbolic: "Museum of the Second World War: Westerplatte or the Battle of Britain."

On April 5, 2017, the Supreme Administrative Court overruled the January decision of the Provincial Administrative Court suspending the merger of the museums. The court decided that the minister's decision to combine museums was an act of internal management and not subject to the purview of administrative

---

[7] Piotr Gliński, Gość Wydarzeń, Polsat News, March 23, 2017, www.polsatnews.pl.
[8] Mikołaj Chrzan, "Muzeum II Wojny Światowej. To Westerplatte czy Bitwa o Anglię," *Gazeta Wyborcza*, March 29, 2017.

courts.⁹ The court thus excused itself from judging whether the minister had acted in accordance with the law or had breached numerous provisions, as per the complaints of the Museum, the ombudsman, and the city of Gdańsk. This ruling contradicted the interpretation previously issued by the Provincial Administrative Court, and many commentators saw it as an instance of "procrastination" as the courts were under increasing pressure from the government of the Law and Justice Party.[10]

In view of the Supreme Administrative Court's decision, Adam Bodnar, the ombudsman for the rights of citizens, lodged a cassation complaint. It reflected the essence of the dispute over the Museum, the meaning of which went far beyond the mere fate of our institution and concerned the very understanding of civil liberties, the autonomy of culture, and the essence of a democratic state of law. Bodnar argued that museums and other cultural institutions are autonomous from their state funder, in this case the minister of culture.

> If it were assumed that the merger or liquidation of a cultural institution is only an internal matter of its organizer, then public authorities, guided by the tastes and views of the people representing them, could unrestrictedly interfere with freedom of artistic creation, freedom of expression or freedom of scientific research, transforming or liquidating cultural institutions that did not match their views, preferences, and tastes. This manner of implementing the law leads to the deprivation of liberty specified in art. 54 par. 1 and art. 73 of the Constitution of the Republic of Poland... . When adopting the concept that the act of transforming or liquidating cultural institutions is an act of internal leadership, public authorities (state or local government) will be able to close any theater exposing politically incorrect art or close to the public any museum exhibition that does not correspond to current historical policy, and its decisions in this respect will not be subject to any external control.

The ombudsman for the rights of citizens also recognized that the court's evasion of the ruling on the issue of the museums' merger violated the "constitutional right to due process."[11]

After the Supreme Administrative Court's decision, things moved very fast. We had no information from the Ministry of Culture on what would happen, which was already the norm. However, the next day at noon, the director of the Department of Cultural Heritage, Paulina Florjanowicz, appeared in the Museum in the company of the new director, thirty-four-year-old Dr. Karol Nawrocki. He was a historian working at the Gdańsk Institute of National Remembrance, dealing mainly with the Solidarity movement in the city of Elbląg and with

---

9 Postanowienie NSA, April 5, 2017, www.nsa.gov.pl.
10 See Ewa Siedlecka, "Prawo i sprawiedliwość dla PiS," *Polityka* 20 (May 17–23, 2017).
11 Skarga Kasacyjna Rzecznika Praw Obywatelskich, June 7, 2017.

the history of football. He had no museum experience. He was mostly known for promoting the cult of "cursed soldiers" and the operations of the Lechia Gdańsk football fan club. That he was close to the Law and Justice government and, above all, to Minister Sellin was no secret in Gdańsk, and his name had been listed on the stock exchange of my "successors" for some time.

Director Florjanowicz, who in previous months had eagerly participated in the war against the Museum, informed me that the Museum had already been formally liquidated, deleted by the minister of culture from the register of museums, and I was no longer the director because the facility I was managing had ceased to exist. This turned out not to be true; it took a few more days to remove the Museum from the register and to enter the new Museum of the Second World War into it, days during which I could still have exercised my function. However, I did not want to resist anymore, to give them an excuse to attack me for not wanting to leave my position.

The seizure of power took place in a great hurry and in chaos. The new institution did not have a tax identification number for a few days, so the sale of tickets had to stop. Downstairs, in the vicinity of the exhibitions, there were crowds of disoriented and increasingly furious people. All this resembled the atmosphere of an inept coup. Contacts with the media were taken over by a new press spokesman who entered the Museum together with Nawrocki. Little was known about him, except that he was an IT specialist and, as he presented himself on right-wing portals, a "homoskeptic" (also a "Catholic, conservative, proud of being a Pole"). In his online posts he unmasked homosexual threats in contemporary mass culture. He saw the propaganda of homosexuality in the *Sherlock* TV series and *Pirates of the Caribbean* movies, where he was concerned with the "effeminate and makeup-wearing Johnny Depp."[12] This was not only a change of personnel but also a deep cultural shift taking place both in the Museum and in Poland as a whole.

Under the labor code the new director was obliged to offer me a new position in the newly created Museum of the Second World War, but both sides knew that I would not accept it. By doing so I would legitimize the disregard (and, in my opinion, obvious violation) of the law by the minister of culture. Needless to say I also did not see myself working as a subordinate of Dr. Nawrocki. He informed me up front that he intended to introduce changes to the exhibitions suggested by historians with "conservative leanings," mentioning the name of Jan

---

12 Krzysztof Katka, "Homosceptyk rzecznikiem Muzeum II Wojny Światowej," *Gazeta Wyborcza*, April 10, 2017.

Żaryn. He admitted that he had to get acquainted with the exhibitions, because he had only had a quick look at them during the presentation on January 23.

On April 7, 2017, I left the Museum and Gdańsk, returning to my home in Warsaw. A day later, Michał Łuczewski, a sociologist at the University of Warsaw and at the same time a well-known intellectual associated with conservative Catholic circles, visited the Museum. Immediately, he wrote to me to convey his impressions. I will quote this letter not only because it is a record of those moments but also because it is a testimony to how an exhibition can be received by a person very far away from me ideologically but also unprejudiced about the Museum:

> Yesterday, it was impossible to get to the Museum—tickets were sold out already in the morning. People stood in long queues. Asking the management for tickets did not work either, because people at higher levels of your management were afraid to give me a ticket to the "old" exhibition because (I felt) it could have been badly received by the new team. You probably know all this, anyway. But I got in! This is the most beautiful museum I've ever seen. I was repeatedly moved to tears, the enormity of this event was splendidly shown. From the Polish perspective. Beautifully (although this is a bad word) and truthfully, the participation of Poland in the annexation of Zaolzie was shown. I was proud that we did not try to tone it down. I understood Jedwabne better than ever—although for years I have read Wassersztajn's testimony with my students. I do not know just how reliable it is, but only one small fragment of it was on the exhibition. The guide whom I joined was spirited and placed the Polish perspective even stronger in the foreground. He said, referring to the paralyzing sentence of von Stauffenberg in a letter to his wife, "At this point you may think that Germans are bad people. Do not think so. Think that each of us can become such a person through various whispers." I could write a lot, but everything comes down to this: you can say *exegi monumentum* [I have made a monument]—for us.

A few days later, Joanna Muszkowska-Penson sent a dramatic appeal to Minister Gliński not to change the exhibitions. "I am appealing to you, Sir, as one of the last survivors of the Second World War," the former prisoner of Ravensbrück wrote.

> I want you to respect the experience of my generation and stop the destructive activities against the Museum of the Second World War in Gdańsk and thus the common good. I've been to the Museum three times, and I can say without hesitation that this is a place that teaches humility and, despite its horrifying messaging, gives hope that good wins over evil. That is why one has to see it! Meanwhile, you, Sir, did not see the exhibitions, but you judge and criticize it... . You assured us, veterans, of your concerns for the shape of Polish memory. Since we are talking about the same issues, I am asking you to let the Museum function in the shape created by Professor Paweł Machcewicz, Dr. Janusz Marszalec, Professor Piotr M. Majewski, Professor Rafał Wnuk, and other people who for years struggled to commemorate the victims of the most terrible of wars... . I also have a very personal reason for that. My friends, inmates from the minors' cell (aged 15–17)

who died in Ravensbrück, executed or after bestial experimental operations, do not have graves today. For me, the Museum was the only place to restore their memory. You have taken this place away from me.[13]

A few days later, Rafał Wnuk was removed from the Museum (he did not accept a new position as an entry-level researcher in the research department that he had created), and Piotr M. Majewski, whom the new director did not even offer further employment, also left. Out of the four people who had made the largest contribution to the creation of the Museum, only Janusz Marszalec accepted a new position. He did this so that, at that crucial moment, he could monitor closely what the new management was doing and in order to stay with the younger employees, for whom the situation was also difficult. New people, including local Law and Justice activists and an assistant to Deputy Minister Sellin, were appointed in the place of the those who had been dismissed.

Piotr Semka published a triumphal article, expressing satisfaction that I had been removed and calling for a quick change to what we had created. "Paweł Machcewicz, the former head of the Museum, treated the outpost as his private farm." Semka also emphasized his merits and dedication in the fight against the Museum. "I know what price is paid for my own opinion," complained the columnist, "because for a year now Paweł Machcewicz has wiped his lips with my name, sometimes offending me in a particularly severe manner. If that was the price to move away from the most shocking plans for the exhibition, it was worth it."[14] Another right-wing publicist, Andrzej Potocki, called on me to publicly apologize to Piotr Semka.

The blessing of the building by the metropolitan archbishop of Gdańsk, Sławoj Leszek Głódź, was a symbolic takeover of the enemy's space and the exorcism of their ghosts. A choir named after the Pope John Paul II performed the popular, joyful songs "War, O My War," "Heart in a Backpack," and "Lancers Came By." It was indeed a completely different understanding of the war than the one that could be read in the exhibitions next door. I followed with interest the comments on the Museum's Facebook page. One read, "Did the archbishop give absolution for stealing the museum from its real creators, for firing them from their jobs, and for this shameful takeover? To hell you will all go, you will."

Archbishop Głódź did not stop at blessing the building, but he referred to the Museum in his homily during the feast of Corpus Christi. He spoke of a "demon of progress and modernity that is wreaking havoc, for example in Western Eu-

---

13 Joanna Muszkowska-Penson, "Odebrał mi Pan miejsce pamięci," *Gazeta Wyborcza*, April 12, 2017.
14 Piotr Semka, "Nowy początek w muzeum," *Do Rzeczy*, April 18, 2017.

rope." He also mentioned crosses disappearing from public institutions, Christophobia, and allergies to Christians and their faith, ethos, culture, and customs. This led him to the issue of our Museum, which "grows out of this trend of progress and modernity." He expressed outrage that it did not take into account the martyrdom of the Catholic Church—which was actually completely wrong.[15]

The speech of the president of the Institute of National Remembrance, which referred to concentration camp prisoners forced to provide sexual services in camp brothels as "prostitutes," was also symbolic of the new sensitivity prevailing at the Museum. *Polityka* quoted extensively from this speech. "Here [in the Museum of the Second World War] there are thousands of tons of concrete, millions of the Polish taxpayer's money, 33 thousand square meters of space, and from these 33 thousand square meters, 15 cm are devoted to Captain Pilecki. And in what context—near the prostitutes from concentration camps. This is how the story of the Second World War is told here. We cannot afford such a story. People from all over the world come here," Jarosław Szarek warned, noting "the scandalous situation, which the previous leadership led by Prof. Paweł Machcewicz allowed to happen."[16]

Changes at the exhibition were announced personally by Piotr Gliński, the deputy prime minister and the minister of culture: "The display will be changed gradually, because in my opinion the way that stories are manipulated there is inadmissible. There is nothing about father Kolbe. Irena Sendler is hidden somewhere behind a hydrant. Captain Pilecki is shown modestly; there is no phrase "cursed soldiers." It is true, as Prof. Żaryn wrote, that there is almost no Polish church there. Poland was and is a Catholic state."[17]

Shortly thereafter, the minister of culture tossed some familiar ideological objections into a letter to the authors of the exhibitions: "It was you, Gentlemen, who created the museum as an ideological project: the so-called 'European version of history.' ... We do not want a second House of European History in Poland."[18] These assessments were completely opposite to the opinion of the vast majority of several hundred thousand visitors who had seen the Museum

---

15 Jarosław Makowski, "Katolicyzm tak, Bóg nie," Newsweek.pl, June 16, 2017.
16 "Skandal w MIIWŚ! Szarek: Tylko 15 cm dla rotmistrza Pileckiego! 'I to w jakim kontekście—nieopodal prostytutek w obozach koncentracyjnych,'" wPolityce.pl, May 28, 2017, https://wpolityce.pl/historia/341744-skandal-w-miiws-szarek-tylko-15-cm-dla-rotmistrza-pileckiego-i-to-w-jakim-kontekscie-nieopodal-prostytutek-w-obozach-koncentracyjnych.
17 "Wałęsa to Myszka Miki wykorzystywana w walce politycznej. Rozmowa Jacka Nizinkiewicza z wicepremierem i ministrem kultury Piotrem Glińskim," *Rzeczpospolita*, September 6, 2017.
18 List ministra kultury i dziedzictwa narodowego Piotra Glińskiego do Pawła Machcewicza, Piotra M. Majewskiego, Janusza Marszalca, Rafała Wnuka, September 19, 2017.

before the minister of culture. Apparently, these were mostly "snitches and common haters," serving not Poland but foreign interests—that was how the Minister of Culture characterized people with views different from his own in an interview.

The first change introduced by the new management was the removal of a display called *There Once Was Wiadrownia*. It talked about the history of the district in which the Museum was created and about the Museum itself. Director Nawrocki removed the multimedia presentation about the circumstances of the Museum's creation, measures taken by the minister of culture against it, and the defense of the autonomy of the institution. To protest this act of censorship, Janusz Marszalec, the last of four people who created the institution from the very beginning, resigned from his position at the Museum.

Even before the "coup" on April 6, I repeatedly declared, with the support of my closest coworkers and cocreators of the exhibitions, that we would defend the integrity of exhibitions on the basis of copyright law. We had authorship of the contents, just like book authors. Minister Gliński could order the removal of all the exhibitions that we created and install completely new ones in their place (for example, displays created by the "reviewers" and the new director). However, to remove some of the elements or even to add new parts without the consent of the authors was a violation of copyright, as well as the historical, intellectual, and artistic coherence of the exhibitions. In the future, of course, new historical facts or technical solutions might justify changes. That, however, would not happen overnight, immediately after the opening of a museum created over eight years; rather it would result from processes that take time. The arbitrary will of the minister and his political party is not a justification. These are not the circumstances in which museums and historical exhibitions exist, being subject rather to completely different laws and rhythms. Their perspectives should be as far away as possible from party politics. The permanent exhibition of the United States Holocaust Memorial Museum in Washington, DC, which opened in 1993, has so far been visited by over 40 million people, and discussions about its updating, including wider portrayal of the perpetrators of the crimes, only began recently.

In April, we made public the legal opinion prepared by an attorney at law, Maciej Ślusarek, one of the most outstanding experts on copyright in Poland, who had agreed to represent the creators of the Museum's exhibitions if there were ever a trial to defend their integrity. Below is a fragment of the legal opinion. Despite all the statements of President Jarosław Kaczyński and other politicians of the ruling party (not to mention Archbishop Głódź), copyright may turn out to be the last barrier against arbitrary, politically motivated changes to the exhibitions.

Ślusarek argued, "Any manifestation of creative activity of an individual nature, determined in any form, regardless of the value, purpose, and manner of expression, is the subject of copyright." In conclusion, he wrote,

> Based on the analysis of regulations and jurisprudence, and then referring to the specifics of the subject of this analysis, which is a permanent exhibition of the Museum of the Second World War in Gdańsk, taking into account the manner in which it was implemented and the complexity of the concept of its creation, developed by its authors, it undoubtedly constitutes a work within the meaning of the Act on Copyright and Related Rights. It should be pointed out that in the present situation, the work is not only the final version of the exhibition in the form of an installed exhibition implemented in cooperation with Tempora but also the script of the exhibition itself, authored by Professor Paweł Machcewicz, PhD, Dr. Piotr M. Majewski, Dr. Janusz Marszalec, and Dr. Rafał Wnuk. These people, as the creators of the exhibition, are entitled to nontransferable copyrights, and any changes to the work can be made only with the consent of all creators.
>
> This means that any interference with the current form of the exhibition ... is a violation of the nontransferable, personal authorship rights of the creators of the exhibition ... and may also be subject to criminal law. The creators have the right to demand the cessation of violations (interference in the exhibition), and if these are made ... they have the right to demand that the exhibition be restored to its original state.[19]

Of course, we do not know how long independent courts will exist in Poland, but there is always a possibility of appealing to European courts. I also believe that sooner or later, the rule of law will be restored in Poland, and those responsible for its violation will be liable, including for arbitrary interference in the shape of the exhibitions. Even if the exhibitions are changed, it will be possible in the future to restore them to their original shape.

The Law and Justice leader Jarosław Kaczyński, who from time to time gave strategic guidance to his camp on how to proceed with actions against the Museum, referred to it again in July 2017. This time he spoke after a very dramatic wave of mass demonstrations on streets of Poland in defense of the independence of the courts. Kaczyński began by stating that the Germans never repaid their obligations to Poland and that they imposed their historical policy on us. "For example, the Museum of the Second World War in Gdańsk, such a special gift of Donald Tusk to Angela Merkel, is nothing else but an inscription into the German historical policy. This is a museum that fits in with the German historical policy. When we want to change this, and when the minister of culture does change it, the ombudsman complains about this to the courts, and the courts

---

19 Opinia prawna mecenasa Macieja Ślusarka, Kancelaria Leśnodorski, Ślusarek i Wspólnicy, April 24, 2017.

order that these changes be withdrawn," the leader of Law and Justice recalled. "This is the situation in Poland today and hence this reform [of the judiciary] to which we keep coming back. Because it is a very important matter, and it is so absolutely necessary."[20]

And so all the threads were finally connected: museums, phobias of other nations, historical politics, and courts. If the latter had already been subordinated to the ruling party, we would have had no chance to open the Museum and present the exhibitions to the public. As long as the Museum remains open and the exhibitions continue essentially as they were originally created, despite those politically motivated changes that were introduced afterward, no lie can harm it. The crowds still keep coming to the Museum, and the waiting list to buy tickets still keeps growing. In less than two months from the opening, 100,000 people came to see the exhibitions; in the first year, more than 600,000 visited. And by the time this book is published, there will have been even more. It is for all these people that the Museum of the Second World War was created, and they are the ones who have the final say in judging it.

---

[20] "Kaczyński: Muzeum II WŚ w Gdańsku wpisuje się w niemiecką politykę historyczną," Polska Agencja Prasowa, July 28, 2017, https://www.pap.pl/aktualnosci/news%2C1027686%2Ckaczynski-muzeum-ii-ws-w-gdansku-wpisuje-sie-w-niemiecka-polityke-historyczna.html.

Not the End of the Story

So, what is the meaning of this story? I think that in disputes surrounding the Museum of the Second World War, there are at least a few fundamental issues: the construction of national identity, the role of history and culture, and understanding our times.

The most frequently repeated accusation against the Museum was that it showed the experiences not only of Poles but also of other nations. We had allegedly created not a Polish but a "European," "cosmopolitan" museum, marked by "pseudo-universalism." If we try to understand what is behind this ideological conjuring, we will see a deep conviction that a mere display of Polish experiences as part of European and world history poses a threat to Polish individuality and uniqueness. We should deal only with ourselves and prove that we have experienced the most difficult fate as a nation and shown the greatest heroism, both unmatched by any other nation. According to this thinking, to present in the museum—which is, after all, devoted to a global conflict—the fates of other countries can only threaten the image of Poland and "blur" it in a wider landscape. This is a vision of Polishness constantly under threat, underlined by uncertainty and fear. For how else can one understand the statement that it is a crime to show, next to the Warsaw Uprising, uprisings in Paris, in Slovakia, and in Prague and that it is a crime to show the Polish Underground State and Polish Home Army next to other conspiracies and resistance movements in occupied Europe? The history of every nation is, in a sense, unique, distinct from the experiences of other nations, and showing them side by side can only serve to better reveal the differences and similarities. If we are afraid of that, we give up any chance of a dialogue, not only to show our history to others but also to understand others' experiences. We lock ourselves in a "besieged fortress"; we lift the drawbridge, fortify the walls, and repeat to ourselves that we are exceptional and the greatest. Only, what is the use of this for the national identity? And why do we expect others to listen to our history if we do not want to hear about their experience?

I think that attacks on the "European" and "international" character of the Museum of the Second World War must be seen as part of the rising wave of isolationism, nationalism, and even xenophobia in recent years. In the Polish context, these are part of the activities that have led to the weakening of our country's ties with Europe and the gradual "withdrawal" of Poland from the European Union, which is more and more often presented not as the personification of the democratic West to which we have aspired over decades of isolation but as a source of danger to our identity. This is not, of course, a purely Polish phenomenon but part of a much broader tendency to reject international cooperation, European integration, globalization, and even a community of universal human values. The same is happening in Hungary under Viktor Orbán; the National Front in France, Alternative for Germany, and supporters of Donald Trump preach similar ideas. "The ideological battle tearing the world apart is perfectly illustrated by the fight around the Polish museum," American author Hanna Kozlowska has noted. "Attempting to suppress the story of the Second World War, which emphasizes its international character, is, in a way, yet another part of the rejection of globalization." In addition to pointing out the parallel between the attacks on the Museum and the mentality of American and European isolationists, Kozlowska adds,

> If you look away from the global character of war, you can't make connections between the war in Syria and what happened to Poland during World War II—which the museum quite literally attempts, showing images of Syrian refugees in a film that concludes the exhibit. It recounts the postwar history of the world, still filled with armed conflict. Once you decide to ignore these links, it's much easier to refuse letting in refugees to the country, something the Polish government has been adamant about. The Museum's current director, appointed by Law and Justice, is not a fan of this film, which he says is confusing.[1]

This was a prophetic comment. Removal of this film was the first change in the permanent exhibition introduced after Law and Justice took control of the Museum.

The attitudes toward the Museum are, of course, not only a modern phenomenon. They are deeply rooted in the Polish past. Historians have described hostility toward "foreignness," especially that coming from the West, as one of the constitutive features of the Polish nobility's mentality in the seventeenth and eighteenth centuries. To the nobles, writes well-known historian Janusz Tazbir, the foreigner "carried with him a threat to such basic values as, on the one hand, the political system and religion and, on the other, their language, customs, and historical tradition—all factors that make up the Polish national consciousness."[2] Poland became not only the "fortress of Christianity" but an island besieged from all sides, surrounded by enemies who wanted to subdue it by imposing foreign paradigms upon it. Cultural exchange, travel, and study abroad were frowned upon. Historian Jerzy Michalski notes,

---

1 Hanna Kozlowska, "History Repeating: The Ideological Battle Dividing the World Is Perfectly Illustrated by This Fight over a Polish Museum," *Quartz*, August 19, 2017.
2 Janusz Tazbir, "Stosunek do obcych w dobie baroku," in *Swojskość i cudzoziemszczyzna w dziejach kultury polskiej*, ed. Zofia Stefanowska (Warszawa: PWN, 1973): 84, 91.

An important feature of so-called Sarmatism[3] was the conviction of Poland's essential otherness from the rest of Europe and a strong belief in this alleged fact. It was not limited to the doctrine of the superiority of Poland and Poles over the rest of the European nations, which was introduced in the 16th century and further developed during the 17th century. It was also expressed in the conviction that the cultures of other European countries might work for them, but they were not appropriate for Poland; only Polish customs were suitable to the Polish national character.[4]

This tension between opening up to other countries and drawing on Western experiences to strengthen Poland, or withdrawing into the "fortress" of Polish separateness, tradition, and homeliness, was one of the axes of culture and mentality throughout the partition period (1772–1918), when Poland disappeared from the word map, divided between Prussia, Russia, and Austria.

Experiencing attacks on the Museum, I often had the impression that this was yet another manifestation of the same conflicts and attitudes that dated back centuries. The similarity was not only in the rhetoric around the Museum, such as one reviewer's statement that the main goal of the exhibitions should be to show "who we are today and why we are the way we are: freedom-loving, Catholics, patriots, etc.; and above all, proud of our history." That would, of course, be too simple an association. The similarity was much deeper: the conviction of one's own monopoly on Polishness and patriotism and the right to exclude from the national community those who understand these values differently. It was also about the vision of a still-threatened Poland, forced to guard at all costs its individuality and uniqueness, which could be weakened even by showing, in one museum, the experiences of other nations that had fought and suffered during the Second World War.

Another striking feature of the attacks on the Museum was the dismissal of "civilians" in the presentation of wartime experiences as somehow inferior, less important than soldiers (whose stories were, of course, also shown in the exhibitions). This perception is extremely surprising in a country that suffered such terrible civilian casualties and where memory of the war is dominated, even today, by the experiences of the civilian population. This was well demonstrated by the extensive sociological research carried out in 2009 by the Museum of the

---

3 Sarmatism was a system of social and political beliefs of a group within the Polish nobility in the seventeenth and eighteenth centuries. Its members called themselves Sarmatians and traced their ancestry to steppe peoples inhabiting Eurasia from the sixth century BC to the fourth century AD. Sarmatism epitomized the most conservative and anti-Western trends among Polish nobility and gentry.
4 Jerzy Michalski, "Sarmatyzm a europeizacja Polski w XVIII wieku," in *Swojskość i cudzoziemszczyzna* (Warszawa: Instytut Badan Literackich PAN, 1971): 114 (conference proceedings).

Second World War, the first such in-depth research conducted in free Poland after the fall of communism. In this sense, the Polish memory of war is completely different from that of, for example, Americans, whose civilians did not die on such a massive scale and where the memories center on the military campaigns in the Pacific and in Europe, as shown in the National World War II Museum in New Orleans.

Also, this disregard for the perspective of the civilian population reflects the well-established conviction in Poland that the military-insurgent part of our tradition is more important, more valuable than other, more "civilian" ways of serving the country and the community. In the nineteenth century, this was a classic dispute between supporters of insurrection and those who opted to improve the economic and social condition of the partitioned nation. Yet, in discussion of the experiences of the last war, talking about such an unambiguous dilemma is difficult. Even in the era of the partition, social attitudes were not bipolar, and veterans of uprisings often became enthusiastic supporters of positivist social and economic work. As for the Second World War and Polish resistance, the most important issue is remembering not only the soldiers but also the phenomenon of self-organization within the Polish Underground State, which was created largely as a result of grassroots initiatives, democratic and politically and ideologically pluralistic, in which several hundred thousand conspirators participated. If we add to this the secret underground education system, the number is even higher. The story of the Polish Underground State is extensively covered in the Museum of the Second World War; it is one of its most important sections. However, this fact did not attract the attention of right-wing critics who repeated a mantra that they wanted to see more about soldiers and about "building character" through the experiences of the war. I think that if the bulk of the exhibition remains essentially in its original shape, millions of people who see it will leave the Museum with a much broader, richer picture of the Polish experience during the war depicted beyond the usual patterns. This could be a cornerstone of a deeper reflection on the past and the present, on the ethos of work for the common good, not only under conditions of war. Deep questions of what the war actually means, the violence and suffering associated with it, and how to understand heroism cannot be expressed in simple, comforting formulas. People who see the exhibitions will want to formulate answers for themselves.

Finally, the third dimension of the dispute over the Museum centers on the unprecedented interference of politicians in the shape and fate of the Museum. The autonomy of history and culture from politics is the essence of the democratic state of law that we have enjoyed in Poland since 1989. The persistent campaign against the Museum of the Second World War conducted by the government and the party in power violates these values and brings us closer to behaviors from the time of the Polish People's Republic or contemporary Russia under Vladimir Putin. In Russia, the authorities dissolved the Perm 36 museum, dedicated to the gulag, and in its place established a new institution with a new exhibition completely opposite to the museum created by former prisoners; their association "memorial" was accused of being a "foreign agent." Democratic Poland has never seen such deep political interference in the sphere of history and such drastic actions taken toward a cultural institution. An announcement by the minister of culture that the election win gives the ruling party a mandate to change historical exhibitions and to eliminate inconvenient museums creates a precedent that may be repeated by every subsequent government to come to power. History and culture become politicized and weaponized, part of a party line.

In my opinion, after what has happened, one can no longer speak of any historical policy conducted in the interest of the Polish state. The government of Law and Justice, with great ferocity and energy certainly worthy of a better cause, tried to prevent the opening of the largest historical museum in Poland. One of the primary goals of this Museum, expressed since its very beginning, was to show Polish history to the world and to engage in the discussion on interpretations of history, including the postwar displacements of, among others, Germans. The Museum of the Second World War was opened to the public before the opening in Berlin of the "Visible Sign" museum, whose focus on the expulsion of Germans caused such anxiety in Poland. Instead of supporting the Gdańsk Museum, the Polish government and the ruling party called it a part of "German historical policy." This is beyond the limits not only of decency but of rational thinking, not to mention action for the common good. From the original concept of historical politics formulated by conservative intellectuals at the beginning of the twenty-first century, only rubble, or rather a caricature, remains. Instead of strengthening the national identity, history is used to divide the nation and destroy internal enemies.

All this also undermines elementary pluralism, without which the democratic dimension does not exist. My reply to critics of the Museum of the Second World War was that they had the right to create other museums in which they could present visions of the war and the experiences of Poles different from those shown in Gdańsk. Certainly, it is a difficult endeavor that requires competence, hard work, and time. Instead they chose a dangerous shortcut. Rather than create their own institutions, they decided to discredit and destroy the work of people with different views, stigmatizing them and even excluding them from the national community. This is, essentially, an authoritarian idea. History museums in Poland have to reflect the vision of the ruling party; for others, there can no longer be a place in the public sphere. Can this attempt at "social engineering" and ideological surgery on the minds of people succeed, with conquest after conquest of museums and, with time, other cultural institutions that display "inappropriate" exhibitions, created by "not our" people?

I seriously doubt it. It did not succeed in Communist Poland, where after 1956 the rulers gave up their consistent attempts at ideological uniformization and allowed limited cultural pluralism—and at times even some freedom. Józef Tejchma, in the 1970s the Communist minister of culture and the deputy prime minister (just like Piotr Gliński today) did allow Andrzej Wajda to direct *Man of Marble*, widely perceived in Poland at the time as a film opposed to the Communist regime, denouncing its evils. Tejchma's comrades considered the film "subversive" and did not allow for normal distribution, but the master copy was not destroyed. The film entered the canon of Polish cinema and attained a legendary reputation. In this context, the zeal for subduing Polish history and culture is something new, unprecedented in post-Communist Poland.

Even if the exhibitions at the Museum of the Second World War are changed at the request of the rulers, hundreds of thousands of people have already seen them, and millions have been following the dispute about the Museum. A grassroots social movement was created in defense of the Museum, formed by people who came to demonstrate at the Museum, hung white and red ribbons, signed petitions, and were active on the Internet. This has never happened before in Poland and probably not in any other country. Although the minister of culture contemptuously calls these people "those on the streets and foreign agents," they are part of Polish society, and the questions asked in the Museum of the Second World War are important to them. The sensitivity and vision of Polishness that they find in the exhibition is close to their own: a Polishness that constantly attempts to understand its own past and identity, which are not incompatible with the experience of other nations; a Polishness that believes it has much to offer other nations but can also draw on the achievements of others. "To Europe, yes, but together with our dead," as Maria Janion titled her book.[5] The dispute over the Museum is a new version of the debate over how to understand Polish history, tradition, and patriotism and how to combine them with belonging to a wider community of humans. These questions were asked by successive generations of Poles, and they made us who we are today. Similar questions may often be asked as well by people of other nationalities and cultures since this story, in its deepest meaning, is not only about Poland. It might happen in many other

---

5 Maria Janion, *Do Europy tak, ale razem z naszymi umarłymi* (Warszawa: Wydawnictwo Sic!, 2000).

countries, especially those in which history has become a vital part of public life and dispute.

The removal from the Museum of the people who created it, and even the change to its exhibitions (temporary as it may be; after all, the governments of any party do not last forever) will not erase all these questions. The understanding of history and culture presented by the Museum cannot be annulled and removed by political and administrative means. It is part of both the Polish legacy and the European identity, which grow as a result of subsequent experiences as well as disputes around them. One of the last acts in this drama, still unfolding, is the history of the Museum of the Second World War. Its story is not over; it will go on as long as there are people for whom the issues raised at the Museum are still important.

Bibliography

Anderson, Gail, ed. *Reinventing the Museum: Historical and Contemporary Perspectives on the Paradigm Shift*. Walnut Creek, CA: AltaMira Press, 2004.
Becker, Manuel. *Geschichtspolitik in der "Berliner Republik." Konzeptionen und Kontroversen*. Wiesbaden: Springer, 2013.
Białoszewski, Miron. *Pamiętnik z powstania warszawskiego*. Warszawa: PiW, 1970.
Davies, Norman. *Europe at War, 1939–1945: No Simple Victory*. London: Pan Macmillan, 2006.
Janion, Maria. *Do Europy tak, ale razem z naszymi umarłymi*. Warszawa: Wydawnictwo Sic!, 2000.
Janion, Maria. *Płacz Generała. Eseje o wojnie*. Warszawa: Wydawnictwo Sic!, 1998.
Kostro, Robert, and Tomasz Merta, eds. *Pamięć i odpowiedzialność*. Kraków: OMP—Centrum Konserwatywne, 2005.
Kwiatkowski, Piotr T., et al. *Między codziennością a wielką historią. Druga Wojna Światowa w pamięci zbiorowej społeczeństwa polskiego*. Gdańsk: Muzeum II Wojny Światowej, 2010.
Linenthal, Edward T. *Preserving Memory: The Struggle to Create America's Holocaust Museum*. New York: Columbia University Press, 2001.
Łubieński, Tomasz. *Ani triumf, ani zgon: Szkice o Powstaniu Warszawskim*. Warszawa: Nowy Świat, 2009.
Murawska-Muthesius, Katarzyna, and Piotr Piotrowski, eds. *From Museum Critique to the Critical Museum*. New York: Routledge, 2016.
Osęka, Piotr. *Syjoniści, inspiratorzy, wichrzyciele. Obraz wroga w propagandzie Marca 1968*. Warszawa: Żydowski Instytut Historyczny, 1999.
Ostow, Robin, ed. *(Re)visualizing National History*. Toronto: University of Toronto Press, 2008.
Piotrowski, Piotr. *Muzeum krytyczne*. Poznań: Dom Wydawniczy REBIS, 2011.
Polityka historyczna. Historycy—politycy—prasa. Konferencja pod honorowym patronatem Jana Nowaka-Jeziorańskiego. Pałac Raczyńskich w Warszawie, 15 December 2004, Warszawa 2005.
Pomian, Krzysztof. *Collectors and Curiosities: Paris and Venice, 1500–1800*. Cambridge, UK: Polity Press, 1990.
Rutkowski, Tadeusz. *Adam Bromberg i "Encyklopedyści." Kartka z dziejów inteligencji w PRL*. Warszawa: Wydawnictwa Uniwersytetu Warszawskiego, 2010.
Snyder, Timothy D. *Bloodlands: Europe Between Hitler and Stalin*. New York: Basic Books, 2010.
Stefanowska, Zofia, ed. *Swojskość i cudzoziemszczyzna w dziejach kultury polskiej*. Warszawa: Wydawnictwo Naukowe, 1973.
Szlendak, Tomasz, ed. *Dziedzictwo w akcji. Rekonstrukcja historyczna jako sposób uczestnictwa w kulturze*. Warszawa: Narodowe Centrum Kultury, 2012.
Williams, Paul. *Memorial Museums: The Global Rush to Commemorate Atrocities*. Oxford: Berg, 2007.
Wolfrum, Edgar. *Geschichtspolitik in der Bundesrepublik. Der Weg zur bundesrepublikanischen Erinnerung 1948–1990*. Darmstadt: Wissenschaftliche Buchgesellschaft, 1999.
Ziębińska-Witek, Anna. *Historia w muzeach. Studium ekspozycji Holokaustu*. Lublin: Wydawnictwo UMCS, 2011.

Ziębińska-Witek, Anna. "Victims of the Holocaust in Museum Exhibitions: New Ways of Representation." *Rocznik Instytutu Europy Środkowo-Wschodniej* 14, no. 1 (2017): 135–152.

Index of Names

Adamowicz, Paweł   2, 40, 55, 107, 111, 125
Anders, Władysław   127
Arabski, Tomasz   10, 40, 42

Baranowska, Marta   68
Barnavi, Elie   52, 78 f.
Bartoszewski, Władysław   20, 76 f., 160
Beneš, Edvard   37
Benoit, Isabelle   52
Berezowska, Maja   68
Białoszewski, Miron   88, 136
Bloomfield, Sara   157
Bodnar, Adam   2, 143, 165
Borejsza, Jerzy   76
Borodziej, Włodzimierz   76, 79
Brudziński, Joachim   99
Bryan, Julien   126
Bryan, Sam   126

Cameron, Duncan F.   90
Choróbski, Jerzy   69
Chrzan, Mikołaj   164
Churchill, Winston   73
Chwalba, Andrzej   76, 79
Chwin, Stefan   51, 136
Cichocki, Jacek   32, 42, 60
Czartoryski, Arkadiusz   117 f., 120 f.

Darski, Łukasz   161
Davies, Norman   77, 79, 113
Dowbor-Muśnicka, Agnieszka   64
Dowbor-Muśnicka Lewandowska, Janina   64
Dowbor-Muśnicki, Józef   64
Duda, Wojciech   9 f., 40, 42 f., 48, 103, 140
Dziuk, Barbara   157 f.

Florjanowicz, Paulina   165 f.
Fortuna, Grzegorz   40
Frei, Norbert   18, 28

Gaeta, Christophe   52
Gawin, Dariusz   17 f., 32, 44
Gawin, Magdalena   151

Geirnaert, Didier   52
Gera, Vanessa   113
Gliński, Piotr   41, 50, 54 f., 58, 67, 100, 104, 106 f., 109–114, 116, 125, 130–132, 138, 140–143, 145–152, 154 f., 159, 161, 163 f., 167, 169 f., 178
Głódź, Sławoj Leszek, archbishop   56, 168, 170
Gmyz, Cezary   17
Gomułka, Władysław   98, 153
Gruszkowski, Wiesław   56
Gutman, Israel   77

Halbersztadt, Jerzy   44, 50, 148
Herbert, Ulrich   77–80
Hilberg, Raul   79
Holzer, Jerzy   76 f.

Janion, Maria   135 f., 178
Janowski, Maciej   107
John Paul II   129, 168

Kaczyński, Jarosław   1 f., 12, 16, 19 f., 49, 97–100, 128, 147, 170–172
Kaczyński, Lech   1, 17, 31, 54, 99, 130
Kafka, Franz   21, 125
Kamiński, Łukasz   43
Kamiński, Mariusz   99
Karaś, Romuald   73
Karski, Jan   120
Kasztelan, Antoni   68
Kochanowski, Jerzy   22
Kominek, Bolesław   53
Komorowski, Bronisław   143
Kopacz, Ewa   60
Kopka, Bogusław   19
Kosiewski, Piotr   161
Kozlowska, Hanna   174
Kruk, Elżbieta   119–121
Krzyżanowska, Olga   68
Kula, Marcin   22
Kur, Tadeusz   122 f.
Kurtyka, Janusz   38, 42

Kwiatkowski, Wojciech 24, 147
Kwieciński, Piotr 152 f.

Libeskind, Daniel 44, 50, 56 f.
Lichocka, Joanna 103, 118
Linenthal, Edward T. 27, 74
Logemann, Daniel 49
Lohman, Jack 55
Łokuciewski, Witold 68
Łubieński, Tomasz 134
Łuczewski, Michał 28, 167

Maczek, Stanisław 71 f., 85, 121 f.
Majewski, Piotr M. 13, 15, 17, 21, 37 f., 44, 46, 49, 51, 63, 75, 102, 129, 143, 167 f., 171
Manikowski, Adam 107
Marszalec, Janusz 37 f., 48 f., 51, 60, 63, 99 f., 102, 104, 106, 129, 140, 167 f., 170 f.
Meller, Marcin 132 f.
Merkel, Angela 10, 171
Michalski, Jerzy 174 f.
Michalski, Łukasz 151
Moczar, Mieczysław 123, 153
Momysz-Uła, Bordżuan 73
Motyka, Grzegorz 23
Mrożek, Sławomir 1
Mularczyk, Arkadiusz 43
Muszkowska-Penson, Joanna 68, 133, 163, 167 f.

Nagorski, Andrew 51
Najder, Zdzisław 69
Nawrocki, Karol 165 f., 170
Niwiński, Piotr 99 f., 115, 125, 127–130, 133
Nowak, Andrzej 25, 89, 139 f, 150
Nowak, Sławomir 39–41

Ogrodnik, Wojciech 106, 109
Ołdakowski, Jan 32, 39, 44, 134
Orbán, Viktor 29, 174
Osęka, Piotr 122 f.
Ostaszewska, Maja 92, 125
Ostrowska, Izabela 160

Pągowski, Andrzej 51
Paul VI 129
Pilecki, Witold 86, 98, 169
Piontkowski, Dariusz 116
Płotnicka, Helena 86
Polian, Pavel 78
Pomian, Krzysztof 15, 52, 76 f., 79
Popek, Leon 70
Pospieszalski, Jan 21, 119, 128
Putin, Vladimir 30, 53, 71, 78, 114, 177

Rousso, Henry 78 f.
Ryszka, Czesław 149

Schetyna, Grzegorz 92
Schmidt, Maria 29
Sellin, Jarosław 39, 41, 50, 54, 100–103, 105–107, 109, 112 f., 116, 118, 125 f., 138, 140 f., 143, 145 f., 148, 150, 155, 158 f., 166, 168
Semka, Piotr 16 f., 21, 53, 78 f., 112, 118, 126–130, 133, 168
Sendler, Irena 89, 169
Skelnik, Julian 73
Ślusarek, Maciej 170 f.
Smith, Martin 74
Snyder, Timothy 77, 113 f., 139 f., 156 f.
Sobecka, Anna 118, 121
Stach, Józef 68, 109
Stachecki, Andrzej 68, 109
Stalin, Joseph 46, 73, 77, 117
Steinbach, Erika 7, 17
Stimmann, Hans 56
Sucharski, Henryk 106
Suska, Łukasz 69
Świat, Jacek 119
Szarek, Jarosław 169
Szarota, Tomasz 15, 76, 78 f.

Tarnawski, Aleksander 133
Taylor, Jacek 149, 155
Tazbir, Janusz 174
Tejchma, Józef 178
Tusk, Donald 8–12, 20, 24, 37, 39–42, 53, 56, 58, 60, 164, 171

Wajda, Andrzej   51f., 92, 151f., 178
Wałęsa, Lech   42, 164, 169
Wassersztajn, Szmul   70, 167
Wawer, Zbigniew   143–145, 150, 159
Wilk-Krzyżanowski, Aleksander   68, 109
Winid, Bogusław   71
Wnuk, Bolesław   64, 69, 85, 114
Wnuk, Jakub   64

Wnuk, Rafał   21f., 37f, 49, 51, 63, 69, 75, 83, 99, 102, 115, 129f., 132, 135, 167–169f., 171
Wójtowicz, Mariusz   105f.
Wolff-Powęska, Anna   76

Zachwatowicz, Krystyna   52, 151
Żaryn, Jan   16, 18f., 21, 25, 119, 121, 126f., 129f., 133, 167, 169
Zdrojewski, Bogdan   42

Photos (by Bartosz Makowski):

The Museum of the Second World War in Gdańsk.

186 —— Photos (by Bartosz Makowski):

The main corridor within the premises of the permanent exhibition. On the left is the room devoted to Soviet communism in the 1920s and 1930s.

The room about Italian Fascism.

188 —— Photos (by Bartosz Makowski):

The room about National Socialism. In its center is Hitler's head by famous sculptor Josef Thorak, found in Gdańsk in 2015.

The section devoted to military campaigns, weapons, and the lives of soldiers and prisoners of war.

Photos (by Bartosz Makowski):

The entrance to the largest section of the permanent exhibition, devoted to the Nazi and Soviet terror, deportations, genocide, and the Holocaust. Behind gigantic letters, one can see a cattle car.

Photos (by Bartosz Makowski): —— 191

The room presenting the German terror against Poles in 1939. Its design evokes the planned, organized character of the Nazi destruction of Polish elites.

Photos (by Bartosz Makowski):

The room about the extermination of Jews in Auschwitz. In the showcases there are objects found on the premises of the camp that had belonged to the victims. Behind them is an artistic installation consisting of empty suitcases.

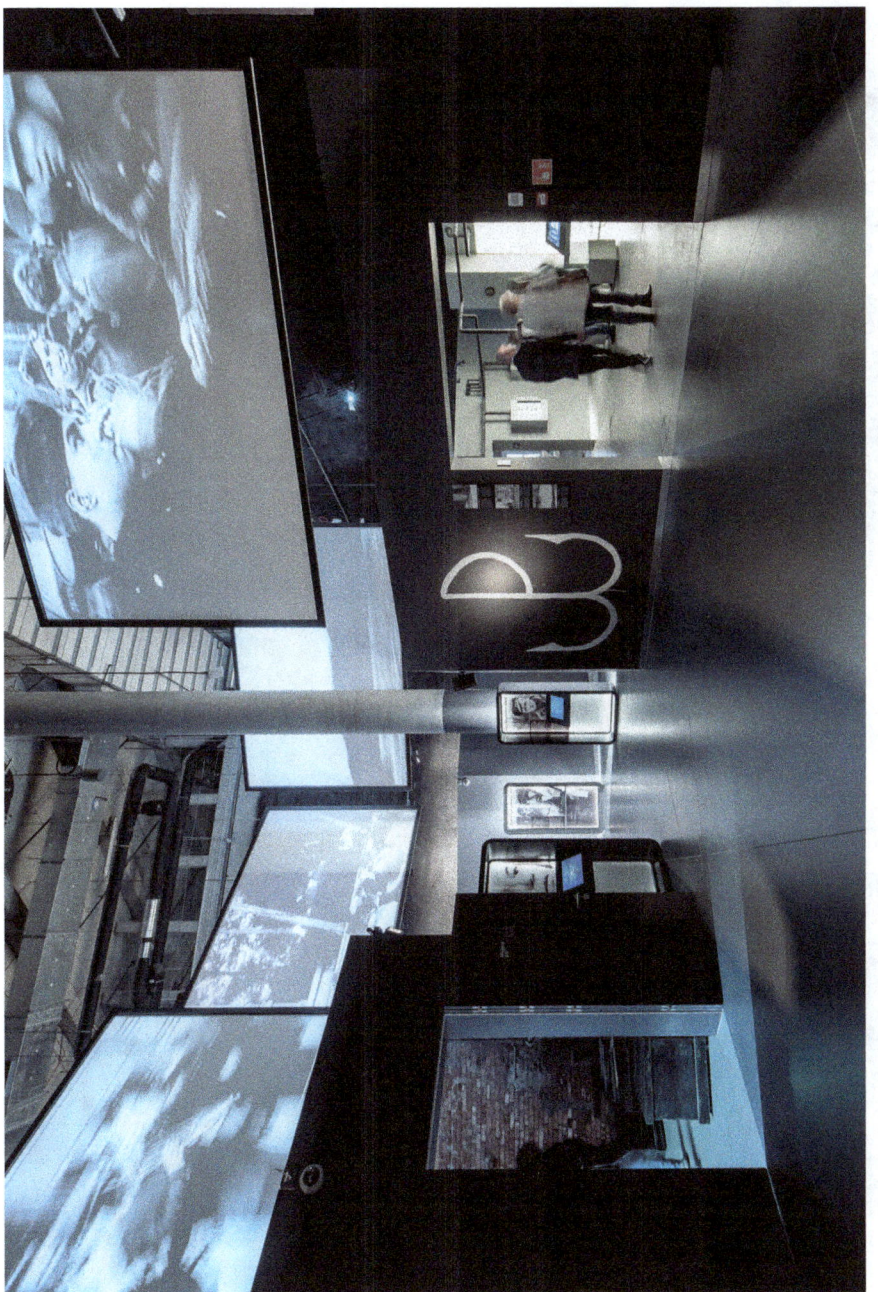

The section of the permanent exhibition about resistance in Poland and other occupied countries. The sign on the wall was an emblem of the Polish Underground State: "Poland fights" (*Polska Walcząca*).

194 — Photos (by Bartosz Makowski):

Europe destroyed by the war. The Soviet tank T-34 has a double symbolic meaning. It signified both the victory over Nazi Germany and the sovietization of East-Central Europe.

www.ingramcontent.com/pod-product-compliance
Lightning Source LLC
Chambersburg PA
CBHW070612170426
43200CB00012B/2667